PRINCETON AERONAUTICAL PAPERBACKS

1. LIQUID PROPELLANT ROCKETS
David Altman, James M. Carter, S. S. Penner, Martin Summerfield.
High Temperature Equilibrium, Expansion Processes, Combustion of Liquid Propellants, The Liquid Propellant Rocket Engine.
196 pages. $2.95

2. SOLID PROPELLANT ROCKETS
Clayton Huggett, C. E. Bartley and Mark M. Mills.
Combustion of Solid Propellants, Solid Propellant Rockets.
176 pages. $2.45

3. GASDYNAMIC DISCONTINUITIES
Wallace D. Hayes. 76 pages. $1.45

4. SMALL PERTURBATION THEORY
W. R. Sears. 72 pages. $1.45

5. HIGHER APPROXIMATIONS IN AERODYNAMIC THEORY. M. J. Lighthill.
156 pages. $1.95

6. HIGH SPEED WING THEORY
Robert T. Jones and Doris Cohen.
248 pages. $2.95

PRINCETON UNIVERSITY PRESS · PRINCETON, N. J.

NUMBER 5
PRINCETON AERONAUTICAL
PAPERBACKS
COLEMAN duP. DONALDSON, GENERAL EDITOR

HIGHER APPROXIMATIONS IN AERODYNAMIC THEORY

BY M. J. LIGHTHILL

PRINCETON, NEW JERSEY
PRINCETON UNIVERSITY PRESS
1960

L. C. Card 60-12052

Printed in the United States of America

HIGH SPEED AERODYNAMICS

AND JET PROPULSION

I. Thermodynamics and Physics of Matter. Editor: F. D. Rossini
II. Combustion Processes. Editors: B. Lewis, R. N. Pease, H. S. Taylor
III. Fundamentals of Gas Dynamics. Editor: H. W. Emmons
IV. Theory of Laminar Flows. Editor: F. K. Moore
V. Turbulent Flows and Heat Transfer. Editor: C. C. Lin
VI. General Theory of High Speed Aerodynamics. Editor: W. R. Sears
VII. Aerodynamic Components of Aircraft at High Speeds. Editors: A. F. Donovan, H. R. Lawrence
VIII. High Speed Problems of Aircraft and Experimental Methods. Editors: A. F. Donovan, H. R. Lawrence, F. Goddard, R. R. Gilruth
IX. Physical Measurements in Gas Dynamics and Combustion. Editors: R. W. Ladenburg, B. Lewis, R. N. Pease, H. S. Taylor
X. Aerodynamics of Turbines and Compressors. Editor: W. R. Hawthorne
XI. Design and Performance of Gas Turbine Power Plants. Editors: W. R. Hawthorne, W. T. Olson
XII. Jet Propulsion Engines. Editor: O. E. Lancaster

PRINCETON, NEW JERSEY
PRINCETON UNIVERSITY PRESS

PREFACE

The favorable response of many engineers and scientists throughout the world to those volumes of the Princeton Series on High Speed Aerodynamics and Jet Propulsion that have already been published has been most gratifying to those of us who have labored to accomplish its completion. As must happen in gathering together a large number of separate contributions from many authors, the general editor's task is brightened occasionally by the receipt of a particularly outstanding manuscript. The receipt of such a manuscript for inclusion in the Princeton Series was always an event which, while extremely gratifying to the editors in one respect, was nevertheless, in certain particular cases, a cause of some concern. In the case of some outstanding manuscripts, namely those which seemed to form a complete and self-sufficient entity within themselves, it seemed a shame to restrict their distribution by their inclusion in one of the large and hence expensive volumes of the Princeton Series.

In the last year or so, both Princeton University Press, as publishers of the Princeton Series, and I, as General Editor, have received many enquiries from persons engaged in research and from professors at some of our leading universities concerning the possibility of making available at paperback prices certain portions of the original series. Among those who actively campaigned for a wider distribution of certain portions of the Princeton Series, special mention should be made of Professor Irving Glassman of Princeton University, who made a number of helpful suggestions concerning those portions of the Series which might be of use to students were the material available at a lower price.

In answer to this demand for a wider distribution of certain portions of the Princeton Series, and because it was felt desirable to introduce the Series to a wider audience, the present Princeton Aeronautical Paperbacks series has been launched. This series will make available in small paperbacked volumes those portions of the larger Princeton Series which it is felt will be most useful to both students and research engineers. It should be pointed out that these paperbacks constitute but a very small part of the original series, the first seven published volumes of which have averaged more than 750 pages per volume.

For the sake of economy, these small books have been prepared by direct reproduction of the text from the original Princeton Series, and no attempt has been made to provide introductory material or to eliminate cross references to other portions of the original volumes. It is hoped that these editorial omissions will be more than offset by the utility and quality of the individual contributions themselves.

Coleman duP. Donaldson, General Editor

PUBLISHER'S NOTE: Other articles from later volumes of the clothbound series, *High Speed Aerodynamics and Jet Propulsion*, may be issued in similar paperback form upon completion of the original series in 1961.

CONTENTS

SECTION E

HIGHER APPROXIMATIONS

M. J. LIGHTHILL

E,1. Introduction. The theories of high speed aerodynamics described in this volume, on the basis of which a quantitative study of the characteristically "nonlinear" effects is made in the present section, are those which neglect the diffusion of temperature by conduction and the diffusion of shear by viscosity. Thus they apply to phenomena in which the regions of boundary layer type, in which the diffusion mechanisms are important, are so thin that the said adiabatic theories give a good approximation to the flow. The influence of the diffusion phenomena may often be sketched in, to improve the theory, after it has been completed. For example, in calculating the drag on a body which moves rapidly through the atmosphere, it is usually necessary to include an estimate of the skin friction due to the laminar or turbulent boundary layer. Frequently, too, at supersonic speeds, it is necessary to take into account the modification of the adiabatic pressure distribution, which results from interaction between a rear shock and the boundary layer.

But the theories which should guide one, in making these final modifications to the adiabatic theory, are described in Vol. IV and V. The present volume is confined to a study of the application of the adiabatic theory to those phenomena in which the diffusion mechanisms are not fundamental, and to the comparison of the resulting predictions with experimental results.

In this restricted class of high speed flow phenomena, the mathematical apparatus necessary to the theory is simplified. The only basic physical concepts which need be considered, besides those already appearing in the theory of the thermodynamics of a gas in equilibrium, are the velocity field at any instant (a concept first used by Euler) and the associated concepts of momentum and kinetic energy. The basic relations, from which all equations used in the theory are deduced, are as follows:

1. The mass of a portion of fluid does not change as its position and shape alter.

2. Its momentum changes at a rate equal to the resultant of the pressure forces acting on it. (Gravity may be neglected in high speed aerodynamics.)

3. Its energy, i.e. the sum of its kinetic and internal energies, changes at a rate equal to that at which pressure forces do work upon it.

4. At any point of the gas, at any time, the pressure, temperature, and the specific thermodynamic variables (specific volume, entropy, internal energy, and enthalpy) are related by the equation of state of the gas. (Thermodynamic relaxation times are unimportant for the flow of *air*, except at very high temperatures, though not necessarily for the high speed flow of other gases such as CO_2 or Freon.)

The theory further postulates that at points not on a shock or a vortex sheet the velocity and the thermodynamic variables are continuous functions. At these points the relations (1) and (2) imply Euler's equations of continuity and momentum, while the relations (3) and (4) imply that the rate of change of specific entropy for a particle of fluid is zero.

A shock, on the adiabatic theory, may be regarded as a (possibly moving) surface of zero thickness. As a particle of fluid crosses a shock, its component of velocity tangential to the shock remains continuous, but the component normal to the shock, with the pressure, temperature and specific thermodynamic variables, all change discontinuously, though subject to relations (1) to (4) above. The changes are fully discussed in III,E; here it is only necessary to remark that the specific entropy increases discontinuously, although it has been seen that except at a shock it cannot change at all. One is still justified, however, in calling the theory adiabatic, since the microscopically thin region, in which the heat exchanges producing this increase in entropy occur, is ignored.

A vortex sheet (otherwise called "contact discontinuity"), on the other hand, in this theory, is a surface of discontinuity through which no fluid passes. Thus the normal component of velocity relative to the sheet is zero on both sides. It follows from (2) above that the pressure is continuous at the vortex sheet; but the tangential component of the velocity, as well as the temperature and the specific thermodynamic variables, are, in general, all discontinuous there.

The relations described above hold throughout the gas, subject to appropriate conditions assumed to hold initially, or at large distances from the phenomena of interest, and subject also to a boundary condition at the surfaces of any solid bodies in the field of flow. This permits a relative velocity between the gas and the portion of body surface in contact with it, but requires that it be tangential to the surface. It is of interest to notice that the discontinuity, between the surface conditions given by the theory and those obtaining in practice, has all the properties of a vortex sheet which were listed above. In fact the theory simply replaces the diffuse vorticity in the boundary layer by a sheet of concentrated vorticity enveloping the body.

The question of existence and uniqueness of solutions to high speed flow problems, if tackled by the adiabatic theory, is discussed in Sec. B. The rather restricted class of exact analytic solutions is fully described in Sec. F. In the present section the study of *approximate* solutions to the equations and boundary conditions of the adiabatic theory, begun in Sec. C and D, is continued. In Sec. C and D every problem was treated by the roughest approximate method which could reasonably be applied. In most cases this consisted mathematically of a complete linearization of the problem. It is these cases which are reconsidered in the present section, to take into account "nonlinear" effects. The two problems studied in Sec. D in which the only applicable approximation was nonlinear, namely, transonic and hypersonic flow, are not treated to any higher approximation in this section.

In a sense, then, this section constitutes a critique of the process of linearizing problems of adiabatic flow. For every approximate theory discussed herein is superior in at least one respect to a completely linearized theory; that is, each is closer to the theoretically exact adiabatic flow; thus they indicate shortcomings of the linearized theories and constitute methods of (at least partially) overcoming these shortcomings. It does not, of course, follow that the improved approximate theories will always be nearer in all respects to the real flow in which the diffusion of shear and heat plays a part. It may happen occasionally that the influence of adiabatic nonlinearity on some measurable quantity is partly canceled out by the influence of the boundary layer. But in the general flow picture the two sources of error distort in quite different ways the simplified version given by the linearized theory. The object of the present section is so to improve the theory that these errors which are due to the nonlinearity of the equations of adiabatic flow are reduced as far as possible.

Expressed physically, the linearization process is always equivalent to assuming that the flow, in some frame of reference (in which the fluid speeds are expected to be relatively small), is governed by the equations of sound. The physical significance of the nonlinearity, which the higher approximations will try to take account of, consists in three differences from acoustic theory, as follows:

5. The velocity of sound is not uniform, but varies with the other thermodynamic variables (for air, in proportion to the square root of the temperature) throughout the flow.

6. At any point disturbances are propagated at the local speed of sound *relative to the local fluid velocity*, not relative to the fixed frame of reference.

7. Because, as a result of (6), in continuous flow compression waves would overtake the waves ahead of them, a discontinuous wave, the shock, must be assumed to appear, always just in time to prevent this happening. The speed of the shock (unlike that of discontinuities in the

acoustic theory) is supersonic relative to the fluid velocity ahead of it, and subsonic relative to the fluid velocity behind it. It can reflect a part of the energy of sound waves which overtake it, and transmit a part of the energy of those it overtakes, and it adds to the entropy of the fluid through which it passes.

It is literally true that the properties (5) and (6) embody all the difference, in the equations of motion for continuous flow, between the exact adiabatic theory and the acoustic theory. For those of the adiabatic theory constitute five linear homogeneous equations in the first derivatives of the pressure, the density, and the three velocity components; the coefficients in the equations are functions of these latter variables. If in these coefficients the pressure and density are replaced by uniform values, with the fluid speed replaced by zero, the equations of sound result. (These, as was seen in C,10, do not *imply irrotationality;* but they state that, if any vorticity is present, it remains where it is independently of any pressure waves which may pass, which in turn are propagated just as if no vorticity were present.) But at any point, of course, the equations of the adiabatic theory are true, with the local values for the coefficients. Hence, locally, propagation is acoustic relative to the local velocity, and with the local value for the speed of sound. This will be the guiding principle throughout the following discussion.

In all the problems treated in this section, the flow is steady in a frame of reference fixed with respect to the solid boundaries. (But of course this is not normally the frame in which, on the linearized small perturbation theory, the equations of sound hold.) We restrict our subject matter thus because on the one hand one-dimensional unsteady motion has been fully treated in III,C, while on the other hand the determination of nonlinear effects in two-dimensional unsteady motion is on the verge of being too complicated a problem for really useful results to be expected in practice, except perhaps in relation to the calculation of flutter derivatives (VII,F). It should also be remembered that, in high speed flow, changes have to be very rapid indeed if there are to be "transient" effects significant compared with the basic transition through a succession of steady states.

In Art. 2 subsonic flow problems are considered. In these there can be no shock and therefore the nonlinearity consists only in items (5) and (6) above. (Actually, the limitation on all the methods used in Art. 2 is not so much the appearance of fluid moving with the local speed of sound, but rather the approach of conditions in which the appearance of a shock becomes necessary.) Now for flow about thin obstacles and similar problems, the small perturbation theory uses (essentially) the equations of sound in a frame which moves with the fluid at large distances from the obstacle. But for subsonic flow about thick obstacles one might expect nearly as good an approximation by using them in a frame fixed with respect to the obstacle. This yields, in steady flow, Laplace's equation. In other words, it gives the same result as is obtained by assuming from the start an infinite velocity of sound, as in the "incompressible fluid" theory. Thus there are two alternative linearizations, one simpler than the other, from which a search for a better approximate solution may begin, which fact leads to a multiplicity of possible approaches to the subject. However, the various improved methods of Art. 2 are all methods of successive approximation which take into account, with increasing accuracy (but in a different order of importance in the various methods), the finiteness of the speed of sound, its variation within the field of flow, and the convection of sound with the fluid.

In Art. 3, et seq., which deal with supersonic problems, the difficulties associated with items (5) and (6) similarly arise, although it will be found that they can be rather more easily surmounted, so that a wider variety of problems may be treated. However, item (7) presents difficulties of an altogether new type. These would make the analytic evaluation of supersonic flows, to really high degrees of approximation, formidable indeed. However, careful study will show, in these articles, that approximations at least better than the small perturbation theory can often be obtained by quite simple processes. Frequently these are based on neglecting those properties of shocks which were collected into the last sentence of item (7).

For the purpose of evaluating most of the quantities of greatest practical importance, the theories to be described merely improve approximations already given by the linearized small perturbation theory. But there are other aspects of the flow as to which information derived from linearized theory is absent, or else completely erroneous, but which can nevertheless be evaluated by the more exact theories. Many of these are apparent from the three points of difference listed above. Almost always, for example, the strength of any shock occurring in the flow falls into this category.

Again, the shape of the solid boundaries may be such as to produce small disturbances on the whole but large disturbances in some particular region (e.g. the rounded leading edge of an airfoil). Then the linearized theory may simply give disturbances tending to infinity in this region, while more exact theories give useful finite approximations to the disturbance distribution.

But there are yet further aspects of flow on which the linearized theory gives quite erroneous information. These may be described generically as "the behavior of any wave after it has traveled a distance of many wavelengths," where the word "many" is to be understood in a sense whose import varies inversely with the amplitude of the wave. For when the wave has traveled such a distance the nonlinear effects (1) and (2), namely, the convection of sound with the fluid and the nonuniformity of its speed—to say nothing of the influence of shocks—may clearly, by their accumulation, have altered the shape of the wave altogether from that given by linearized theory. This fact is relevant to the steady supersonic flow past an obstacle if the flow at some distance from the flight path is required, since this consists of a wave (headed by a shock, and containing at least one more shock farther back) which is propagated diagonally outward from the flight path. The change of wave form due to effects (1) and (2) is found to be of the nature of a spreading out; at the same time the shocks decay in strength by interaction with the wave, whence the problem is often referred to as that of determining the decay of shocks.

Mathematically, the failure of linearized theory in such a case is explained by the fact that an approximation to a partial differential equation, while yielding adequate results in a limited region, may yield a worse and worse approximation to the *solution* farther and farther from where the boundary conditions determining the solution were applied. There is an interesting parallel in Stokes' theory of flow at low Reynolds number past a solid sphere (IV,B). The inertial stresses, being small compared with the viscous stresses, are neglected. The solution which can then be obtained gives a correct first approximation to the surface pressure distribution, and hence the famous drag formula. But the flow far from the sphere is given quite incorrectly by the theory, on which it is symmetric fore and aft of the sphere, from which one could deduce (by simple momentum considerations) the absence of drag. (The general position of approximate theories of supersonic aerodynamics will be found to be similar, e.g. in Art. 4.) The deficiency may be removed by using Oseen's alternative approximation (IV,B) to the equations of motion at low Reynolds number, which is equally accurate near the sphere and much closer far away from it. The solution of Oseen's equations gives the same first approximation to the surface pressures as that of Stokes's but also gives a correct approximate behavior (involving a wake) at large distances from the sphere.

Often the true behavior of high speed flow, in cases such as those discussed above (for example near a shock, or in a small region of large disturbance on the body surface, or in a wave which has traveled a large distance), cannot be calculated by the kind of process, for obtaining further approximations beyond the small perturbation theory, which seems mathematically most natural. In fact, such a process often diverges in such regions, although converging elsewhere, thus making the local failure of the first approximation hardly surprising.

In the present section, on the other hand, some attempt will be made to show that methods based on sound physical thinking, and especially on the principles (5), (6), and (7) above, with the commentaries following them, will not usually lead one into such difficulties. In fact, such physical approaches are used throughout, except in certain cases where it is convenient, as leading the reader on from the simpler to the more complex, to develop a subject historically. This mode of presentation was adopted as giving a unifying theme which would be generally intelligible, and as holding out most hope for stimulating research into extensions of the theory.

But it should be remarked that there exists, as an alternative, a mathematical approach to these problems, which constitutes a semi-mechanical process of ferreting out the answer in any case of this kind in which physical intuition (though perhaps only temporarily) fails one. This is the author's "technique for rendering approximate solutions to physical problems uniformly valid" [*31*], which has been applied successfully to problems of all of the three types mentioned at the beginning of the second last paragraph. The principle of this technique may be described quite briefly as follows (though the fact that it works, and the practical following out of the principle, are both more complicated). At the same time as successive approximations are made to the dependent variables, they must also be made to one of the *independent* variables or to a suitably chosen combination of them, say x. If y is representative of the other independent variables, then each successive approximation, both to the dependent variables and to x, is a function of y and of a new variable z. The first approximation to x is taken as z itself. Now if x were taken identical with z (so that, effectively, there were no transformation of independent variables), then in the "difficult" region successive approximations would diverge. Hence the successive approximations to x as a function of y and z must be so chosen that those to the dependent variables no longer diverge, as far as those which are treated indicate, anywhere in the region of physical interest. For rendering the initial linearized approximation uniformly valid it is necessary only to proceed to the second approximation to x as a function of y and z. (This may be determined so that the second approximations to the dependent variables are not diverging, even in the difficult region, without actual calculation of these second approximations.) To obtain a uniformly valid *second* approximation to the dependent variables, x would have to be evaluated to the *third* approximation. The interpretation of the final solution is that the expression for x as a function of y and z *defines* the new variable z in terms of the physically significant variables x and y, whence the expressions for the dependent variables in terms of x and z become physically significant. The difference between x and z is only significant near the difficult region.

This method, which, for reasons explained two paragraphs back, will *not* be employed in the present section, is explained in full detail in the reference cited above, and further exemplified in papers by Whitham [*32,66*], Lighthill [*33,56*], Legras [*57,58*], and Tan [*59*]. This mathematical approach will occasionally be referred to below as confirming results based on other points of view.

We may close this article with remarks on the chronology of the material presented in Sec. E. The work of Art. 2 and 3 may fairly be called "classical," if account is taken of the relative novelty of all the subject matter of the series. On the other hand, the material presented in Art. 4 to 8 is taken partly from work published in 1948–1950, and even more from work published after February 1951. This shows how rapidly the subject is expanding, and holds out hopes that its ultimate extent and utility will be very great.

E,2. Subsonic Flows. This article is concerned with the steady adiabatic flow of a uniform subsonic stream past an obstacle not thin enough for the compressibility effect to be estimated satisfactorily from the Prandtl-Glauert small perturbation theory. For a given shape of obstacle (relative to the direction of the stream), any two such flows (with different body sizes, and different values at infinity of the velocity and thermodynamic variables) are of course similar if the Mach number $M_\infty = q_\infty/a_\infty$ of the uniform stream is the same in both. The methods of this article for studying the variation of the flow pattern with M_∞ are not applicable for all values of M_∞ less than 1, but evidence will be presented that they are applicable, in principle, for all values of M_∞ less than that value for which a continuous flow without a shock is no longer possible. It has been seen (A,7 and D,33) that this value of M_∞ is less than 1 by an amount which increases with the thickness ratio of the obstacle, at first like the two-thirds power thereof. However, it exceeds the Mach number at which local supersonic speeds first appear, and it will be shown below that the methods given can evaluate certain continuous transonic flows with reasonable accuracy.

The methods work by evaluating a sequence of successive approximations to the exact adiabatic flow, starting either from the "incompressible" flow with $M_\infty = 0$ or from the Prandtl-Glauert approximation to the flow. For the methods to be of value, the sequence must converge in some sense, and, since each approximate flow pattern is continuous, it is fairly evident that there will be no such convergence when the exact flow contains a shock. But when there is convergence it may often be quite adequate to use only the second approximation, and this is usually done; indeed, the calculation of higher approximations than this is in most cases prohibitively lengthy. At the end of the article the results of the methods are compared with those of the Kármán-Tsien method to be described in F,5.

Expansion in powers of the Mach number. A type of approach by successive approximations, which is natural mathematically, though (as suggested in Art. 1) not necessarily the best for many of the physical problems with which we have to deal, consists in evaluating the successive partial sums in a series expansion of, say, the velocity potential ϕ in powers of some small parameter. The earliest approach by successive approximations, made independently by Janzen [*1*] and Rayleigh [*2*], used just such an expansion of ϕ, namely, one in powers of M_∞. The first of their approximations, consisting of only the first term in the series, is the potential in the limit of zero Mach number, when the problem is linear and governed by Laplace's equation. The second approximation embodies the earliest appearance of compressibility effects, as M_∞ increases, and so on. In this approach there is clearly no restriction on the thickness of the obstacle.

It follows from Kelvin's circulation theorem that, in the steady continuous adiabatic flow of a uniform stream past an obstacle, the said velocity potential ϕ exists. Also, from the equation of continuity, and from Bernoulli's equation, with a constant value for the adiabatic index γ, the potential ϕ satisfies a partial differential equation

$$[a_\infty^2 + \tfrac{1}{2}(\gamma - 1)(q_\infty^2 - q^2)]\nabla^2\phi = \nabla\phi \cdot \nabla(\tfrac{1}{2}q^2) \tag{2-1}$$

where $q^2 = (\nabla\phi)^2$ is the square of the fluid speed. Here, the coefficient of $\nabla^2\phi$ is the square of the local speed of sound, and so embodies the difference from acoustic theory (5) of Art. 1. The right-hand side is what would become $\partial^2\phi/\partial t^2$ in axes moving with the local fluid velocity (remembering that the flow is steady in the fixed frame of reference). Thus it embodies the difference (6), i.e. the convection of sound.

Eq. 2-1 suggested the form of expansion used by Janzen and Rayleigh, namely

$$\phi = q_\infty[\phi_0 + M_\infty^2\phi_1 + M_\infty^4\phi_2 + \cdots] \tag{2-2}$$

Here ϕ_0, ϕ_1, ϕ_2, . . . are taken to be independent of the Mach number $M_\infty = q_\infty/a_\infty$ as well as of the main stream speed q_∞ itself.

Substitution of the series (2-2) in (2-1) gives

$$\begin{aligned}\{1 + \tfrac{1}{2}(\gamma - 1)M_\infty^2[1 - (\nabla\phi_0 + M_\infty^2\nabla\phi_1 + \cdots)^2]\}\\ (\nabla^2\phi_0 + M_\infty^2\nabla^2\phi_1 + M_\infty^4\nabla^2\phi_2 + \cdots)\\ = M_\infty^2(\nabla\phi_0 + M_\infty^2\nabla\phi_1 + \cdots)\cdot\nabla[\tfrac{1}{2}(\nabla\phi_0 + M_\infty^2\nabla\phi_1 + \cdots)^2]\end{aligned} \tag{2-3}$$

By expanding each side of Eq. 2-3 into a regular power series in M_∞^2, and equating successively the coefficients of 1, M_∞^2, M_∞^4, . . . in these two power series, the equations

$$\nabla^2\phi_0 = 0 \tag{2-4}$$

$$\nabla^2\phi_1 = \nabla\phi_0 \cdot \nabla[\tfrac{1}{2}(\nabla\phi_0)^2] \tag{2-5}$$

$$\begin{aligned}\nabla^2\phi_2 = \tfrac{1}{2}(\gamma - 1)[(\nabla\phi_0)^2 - 1]\nabla^2\phi_1\\ + \nabla\phi_1 \cdot \nabla[\tfrac{1}{2}(\nabla\phi_0)^2] + \nabla\phi_0 \cdot \nabla(\nabla\phi_0 \cdot \nabla\phi_1)\end{aligned} \tag{2-6}$$

. . . .

result. By Eq. 2-4 it is verified mathematically that the "incompressible" flow is the limit of the compressible flow for infinite velocity of sound.

Since the boundary condition of zero normal velocity at the surface of the obstacle is independent of M_∞, all the coefficients $\phi_0, \phi_1, \phi_2, \ldots$ in the expansion (Eq. 2-2) for ϕ must have zero normal derivative at this surface. Since also, at large distances from the obstacle, $\phi - q_\infty x$ must have a vanishingly small gradient (where the x axis is chosen in the direction of the undisturbed stream), the conditions at infinity on $\phi_0, \phi_1, \phi_2, \ldots$ must be that the gradients of $\phi_0 - x, \phi_1, \phi_2, \ldots$ all tend to zero.

The solution of Eq. 2-4, 2-5, 2-6, . . . , which are of Poisson's type, under these boundary conditions, presents no difficulty of principle if the body shape is one for which a Green's function is known. For two-dimensional shapes this means that the conformal mapping onto a circle is known. But the method is very laborious except for the two simplest shapes, the circular cylinder and sphere, and no attempt has been made for other shapes than these to calculate the potential more accurately than to the second approximation $\phi = q_\infty(\phi_0 + M_\infty^2\phi_1)$. Therefore, although of course the flow past circular cylinders and spheres bears no resemblance to the adiabatic flow unless special methods of boundary layer control are applied, there is valuable information to be derived from the explicit formulas available in these two cases. Firstly, light may be shed on the question of convergence of the Janzen-Rayleigh process and similar processes. Secondly, the order of magnitude and general behavior of the compressibility correction may be expected to be similar for the real flow around bluff bodies with substantial wakes.

For the circular cylinder the formulas

$$\phi_0 = (r + r^{-1}) \cos \theta$$

$$\phi_1 = (\tfrac{13}{12}r^{-1} - \tfrac{1}{2}r^{-3} + \tfrac{1}{12}r^{-5}) \cos \theta + (-\tfrac{1}{4}r^{-1} + \tfrac{1}{12}r^{-3}) \cos 3\theta \quad (2\text{-}7)$$

where (r, θ) are polar coordinates with the center of the circle as origin and the direction of the undisturbed stream as axis, and the unit of length is the radius of the cylinder, were given by Janzen [*1*]. The values of ϕ_2 and ϕ_3 are given in a survey paper by Eser [*3*], ϕ_2 having been first calculated correctly by Imai [*4*] (and incorrectly by other writers who subsequently agreed with Imai) and ϕ_3 independently by Eser and Imai. It is sufficient here to state that the maximum compressibility correction to the velocity field is at the two symmetrically placed positions of maximum velocity, where the speed is

$$q_\infty\{2 + 1.17M_\infty^2 + [2.34 + 0.59(\gamma - 1)]M_\infty^4 + [6.06 + 3.44(\gamma - 1) + 0.50(\gamma - 1)^2]M_\infty^6 + \cdots\} \quad (2\text{-}8)$$

(In Eq. 2-8 the terms in $(\gamma - 1)$ are those due to the variability of the speed of sound; the others are those due to the convection of sound waves with the fluid (as well as to the finiteness of the speed of sound, which Laplace's equation neglects), corresponding to the first and second points of difference between the exact adiabatic theory and the theory of sound which were noted in Art. 1.) On the other hand, near the stagnation points, the compressibility correction reduces the velocity by a factor $1 - 0.42M_\infty^2 - \cdots$.

For $M_\infty = 0.4$, with $\gamma = 1.40$ (its value for air), the four displayed terms of the series within braces in Eq. 2-8 are $2 + 0.187 + 0.066 + 0.031 = 2.284$; the rate of fall of these terms indicates convergence. The sum to infinity is unlikely to be short of 2.318, the value of q/q_∞ which, under these conditions, indicates a local Mach number of 1. Roughly, we may say then that for flow at $M_\infty = 0.4$ past a circular cylinder the Janzen-Rayleigh process converges although the critical velocity is just attained at the ends of the diameter perpendicular to the flow.

For a value of M_∞ rather greater than this the process presumably starts to diverge. A closer estimate of this value emerges from the method of Taylor and Sharman discussed below.

For the flow past a sphere the formulas

$$\phi_0 = (r + \tfrac{1}{2}r^{-2}) \cos \theta$$

$$\phi_1 = (\tfrac{1}{8}r^{-2} - \tfrac{1}{5}r^{-5} + \tfrac{1}{24}r^{-8})P_1(\cos \theta) + (-\tfrac{3}{10}r^{-2} + \tfrac{27}{55}r^{-4} - \tfrac{3}{10}r^{-5} + \tfrac{3}{176}r^{-8})P_3(\cos \theta) \quad (2\text{-}9)$$

where (r, θ) are spherical polar coordinates with the center of the sphere as origin and the direction of the undisturbed stream as axis, and the unit of length is the radius of the sphere, were given by Rayleigh [*2*]. The value of ϕ_2 was found by Tamada [*5*] and is quoted by Eser [*3*]. The maximum compressibility correction occurs (as for the cylinder) at the position of maximum velocity, where the speed is

$$q_\infty\{1.5 + 0.313M_\infty^2 + [0.330 + 0.062(\gamma - 1)]M_\infty^4 + \cdots\} \quad (2\text{-}10)$$

On the other hand, near the stagnation points, the compressibility correction reduces the velocity, by a factor $1 - 0.25M_\infty^2 - \cdots$.

For $M_\infty = 0.55$, with $\gamma = 1.40$, the three displayed terms of the series within braces in Eq. 2-10 are $1.5 + 0.095 + 0.033$. The series probably converges to a sum not far short of 1.71, the value of q/q_∞ which, under these conditions, indicates a local Mach number of 1. As for the circular cylinder, there is probably a range of values of M_∞ exceeding 0.55 in which the Janzen-Rayleigh process converges to a continuous transonic flow pattern. It is reasonable to expect (see below) that this is bounded above by the Mach number at which shock-free flow becomes impossible.

A difference, in the order of magnitude of the compressibility correction to the flow past bluff bodies, between two-dimensional and axially symmetric flow patterns, has emerged. For the circular cylinder the correction factor to be applied to the surface velocity distribution varied between limits roughly $1 \pm \frac{1}{2}M_\infty^2$ for small M_∞. Since this is in agreement with the value $(1 - M_\infty^2)^{-\frac{1}{2}} = 1 + \frac{1}{2}M_\infty^2 + \cdots$ given by the small perturbation theory for thin airfoils, a compressibility correction of this order may be regarded as the rule in two-dimensional problems. On the other hand, for a sphere the correction factor varies between limits roughly $1 \pm \frac{1}{4}M_\infty^2$. Thus compressibility effects are considerably smaller for bluff bodies of revolution than for corresponding two-dimensional bodies, and in particular the value of M_∞ for which local supersonic speeds first appear is higher (e.g. 0.55 for the sphere as against 0.40 for the circular cylinder). But for bodies of revolution the small perturbation theory (C,5) is more complicated than it is for airfoils and a direct comparison with the correction for a sphere is in consequence not possible.

Both Janzen [*1*] and Rayleigh [*2*] attached a further significance to their calculations, which has not so far been mentioned. They observed that the flow pattern in both these problems was symmetrical about a plane perpendicular to the stream, not only in the first approximation ($M_\infty = 0$) but also in the second, which they had calculated, and indeed in all higher approximations, as they were able to show by induction. Hence, for values of M_∞ for which the process converges, the fluid pressures on the front of the obstacle are canceled out by those behind and there is no drag. Thus d'Alembert's so-called paradox, that, in a steady continuous flow of an inviscid fluid past a body, the body can experience no resultant force in the direction of the stream, is extended to compressible fluids.

Here, the essential discovery of Janzen and Rayleigh was that a steady continuous flow existed for some range of values of M_∞. For, this being so, then by considering the rate of change of momentum of the fluid lying (at one instant) between the body and a surface Σ at a large distance, the drag is seen to be

$$-\int_\Sigma p\,dydz - \int_\Sigma \rho(u - q_\infty)(u\,dydz + v\,dzdx + w\,dxdy) \qquad (2\text{-}11)$$

and it follows from Bernoulli's equation in the form

$$\frac{dp}{\rho} + \frac{1}{2}d(u^2 + v^2 + w^2) = \text{const}$$

that the only terms inside the integrals in Eq. 2-11 which are not at least quadratic in the disturbance velocities $u - q_\infty$, v, and w cancel one another out. But the disturbance velocities must fall off sufficiently rapidly at large distances from the obstacle for the integral of their squares over a large surface to tend to zero, by the properties of harmonic functions (for since, at large distances, the disturbances are small, they satisfy the equation of small perturbations, which is Laplace's equation trivially transformed). Hence the drag is zero.

But the existence of such a flow was not obvious. It might have been conjectured that, when a body moves at an appreciable subsonic Mach number, a steady flow is never set up, but the body continues to send sound pulses to infinity. The radiation of this energy to infinity would only be possible if work were done by the body against a drag. The flow pattern would not be symmetrical about any plane perpendicular to the stream, since the source of sound is moving. But, although at supersonic speeds a drag due to this cause does appear (Art. 4), the methods of the present article show that for values of M_∞ below a certain critical value the disturbance set up by the obstacle constitutes a standing wave of constant energy. Of course, as the body is accelerated from rest to its final uniform motion, work must be done to set up this standing wave; this is the work associated with the concept of "virtual mass." Also, if the acceleration is large (i.e. of the order of the square of the velocity of sound divided by the diameter of the body) a comparable amount of additional work is done in sending out a sound pulse [*60,61*]; but this pulse ceases to be emitted when the body speed becomes uniform.

It remains, as far as the Janzen-Rayleigh method is concerned, to mention that for thinner shapes, for which the adiabatic flow is realized more closely in experiment, only two-dimensional flow patterns have been evaluated, and those only to the second approximation $\phi = q_\infty(\phi_0 + M_\infty^2\phi_1)$. The shapes treated were the ellipse and the symmetrical Joukowsky airfoil (Kaplan [*6*] and others quoted by Eser [*3*], who however gives the formulas for the ellipse incorrectly). But since with these restrictions, at least in the absence of circulation, the method is inferior to the simpler Kármán-Tsien method (see below) it is not necessary to discuss these results, or the general formulation [*7*] of the solution for ϕ_1, in two-dimensional problems without circulation. However, passing to problems with circulation, some light is thrown on the influence of compressibility on lift curve slope, for airfoils of varying thickness, by calculations of Tomotika and Umemoto [*8*] on the flow at incidence past a Joukowsky airfoil with circulation fixed by the Kutta-Joukowsky condition. Expanding lift/(lift)$_{M_\infty=0}$ in a series $1 + \frac{1}{2}M_\infty^2 A + \cdots$, the coefficient A is graphed against the thickness ratio t of the airfoil, for the three angles of attack 5°, 10°, and 15°, in Fig. E,2a. On the small perturbation theory A is 1. Thus the figure illustrates the addition to the compressibility effect on lift due to large disturbances, a subject further discussed below.

Any future work on the Janzen-Rayleigh method would advantageously be confined to three-dimensional problems since, while several other methods for tackling two-dimensional problems exist (as will be seen) the effect of compressibility in other classes of problems, including even axisymmetrical flow at Mach numbers below that at which a shock appears, is known only very sketchily indeed. (Note added in proof: A first step in this direction has been taken by Longhorn [62].)

Successive approximation to the density field. The first attempt to improve on the cumbersome Janzen-Rayleigh method, in the case of two-dimensional flow, was made by Taylor and Sharman [9]. Their procedure for deriving successive approximations to the flow pattern has a

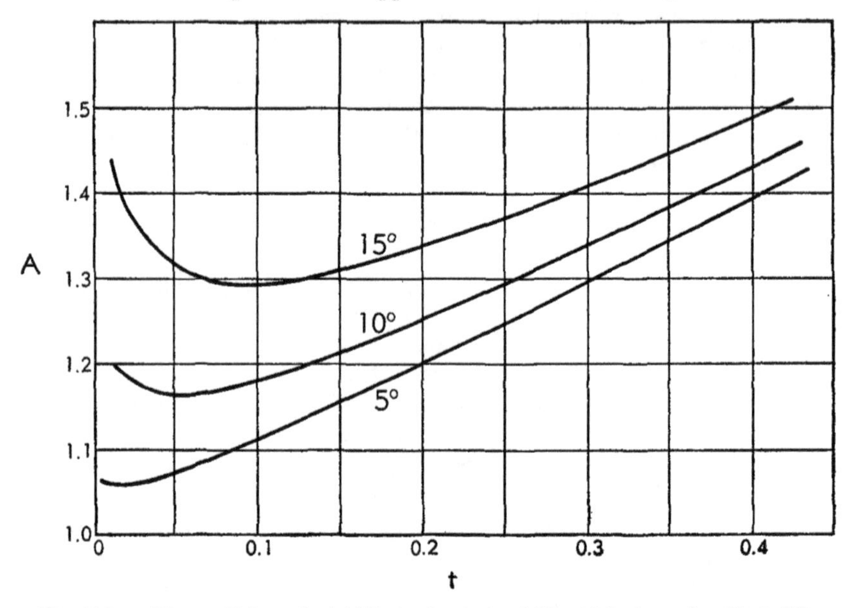

Fig. E,2a. The coefficient A of $\frac{1}{4}M_\infty^2$, in the ratio of lift at Mach number M_∞ to lift at zero Mach number, for a symmetrical Joukowsky airfoil of thickness ratio t at three angles of attack.

more direct physical significance. They work not with Eq. 2-1 for the potential but with the fundamental equations of continuity, irrotationality, and energy conservation

$$\frac{\partial(\rho u)}{\partial x} + \frac{\partial(\rho v)}{\partial y} = 0, \qquad \frac{\partial v}{\partial x} - \frac{\partial u}{\partial y} = 0 \tag{2-12}$$

$$\rho = \rho_\infty \left[1 + \frac{1}{2}(\gamma - 1)\frac{q_\infty^2 - q^2}{a_\infty^2}\right]^{1/(\gamma-1)} \tag{2-13}$$

The idea is to guess a rough approximation to the density field, and, using this value for ρ, to solve Eq. 2-12 for u and v under the given boundary conditions. Eq. 2-13 then (with q derived from these approximate values of u and v) yields a better approximation for the density field. Eq. 2-12 can be solved again using this new value for ρ, and so on. If the process converges, the limit must be the steady adiabatic flow.

The solution of Eq. 2-12 may be accomplished by means of a tank of variable depth, small compared with its length and breadth, and filled with an electrolyte of uniform conductivity. Then Eq. 2-12 holds if (u, v) is proportional to the electric intensity field and ρ is proportional to the depth (analogy A). The obstacle should then be represented by a solid insulator immersed in the electrolyte, and the main stream by a uniform current flow (parallel to it) through the tank, which the obstacle disturbs exactly as it does the fluid stream in the aerodynamic problem.

Alternatively, Eq. 2-12 holds if $(\rho v, -\rho u)$ is proportional to the electric intensity field and ρ^{-1} is proportional to the depth (analogy B). The obstacle should then be represented by a perfect conductor, and the main stream by a uniform current flow perpendicular to it. This second analogy method of calculation, though less natural than analogy A, has the advantage that it is possible to obtain a flow with circulation. For it is easily seen that the circulation

$$\oint (udx + vdy)$$

about the body is directly proportional to the current (if any) supplied by the conductor representing the body, and a nonzero value of this quantity can therefore be obtained by connecting the said conductor, through an appropriate resistance, to one terminal of the battery which generates the current flow.

It should be noticed that in both analogies it is sufficient at each stage to measure only the magnitude of the intensity field at various points, for the purpose of deducing the next approximation to the density distribution. A detailed treatment of the use of the electrolytic tank as a tool in aeronautical research is given in IX, H,2.

Taylor and Sharman always started with uniform depth of electrolyte, corresponding to the approximation $\rho = \rho_\infty$. Their first approximation to the velocity field is therefore the same as that of Janzen and Rayleigh, namely, the incompressible pattern. Subsequent approximations are not identical in the two methods; however, the order of magnitude of the error should be the same at parallel stages. Thus the error in taking $\rho/\rho_\infty = 1$ is of order M_∞^2, and so it may be inferred that (u, v) is given by the first approximation with an error of order M_∞^2. The use of these values in Eq. 2-13 now gives ρ/ρ_∞ with an error of order M_∞^4, giving a second approximation for (u, v) after Eq. 2-12 has been solved, with an error of order M_∞^4, and so on.

However the process just described makes it physically much clearer why there is an exactly steady continuous adiabatic flow (at any rate for values of M_∞ small enough for the process to converge), a point whose significance was explained above.

It also has the practical advantage over the Janzen-Rayleigh method that it can be applied with equal ease to any cylindrical body, including airfoils. Further, although the complete step of measuring the electric intensity field, calculating the density field from it by Eq. 2-13, and reshaping the bottom of the tank accordingly, is laborious, the subsequent repetitions of this process are only equally laborious, while in the Janzen-Rayleigh method they are progressively more so. One result of the latter advantage is that a sufficiently large number of terms can be calculated to make it possible to determine with much greater certainty between convergence and divergence. Thus, for the circular cylinder, the process converged for $M_\infty = 0.4$ and diverged slowly for $M_\infty = 0.5$, and by studying the behavior of the approximations Taylor was able to estimate that the radius of convergence was at about $M_\infty = 0.45$.

There is not perfect accord between the measurements of Taylor and Sharman and the subsequent calculations of the higher Janzen-Rayleigh approximations which were quoted above. The second approximations agree fairly closely, as Taylor and Sharman pointed out, but the convergence for $M_\infty = 0.4$ is rather surprisingly more rapid in their method, in which the third approximation is found to differ little from the second. Some improvement in rate of convergence would be expected since the successive approximations are chosen on sound physical, rather than arbitrary mathematical, grounds. But the velocity near the surface is in consequence given smaller than in the higher Janzen-Rayleigh approximations. The agreement is excellent at a distance of half a radius or more from the surface, taking into account the slight constraining effects of the walls of the tank, each of which measured about nine cylinder diameters. But it gradually falls off as the surface is approached, so that the maximum surface value for q/q_∞, obtained by extrapolation from values measured off the surface, is 2.20, whereas the calculations quoted above from Eser [*3*] indicate a value of at least 2.30 for this quantity. The discrepancy is probably partly due to the large gradients of velocity in this neighborhood which must have made the measurements difficult even though great care was taken. Also, the bottom of the tank had a 15 per cent gradient at this point, so that the accuracy with which the electric field is two-dimensional and satisfies Eq. 2-12 must have been considerably reduced. However, even on the measured values, there is extremely rapid convergence although a Mach number of 0.95 is attained in the field, and the existence of convergence even in transonic flows can, in consequence, hardly be doubted. It would be plausible, in any case, to claim this from the fact that the method is found to converge when $M_\infty = 0.4$, for which it is now believed, from the Janzen-Rayleigh method, that the critical velocity is just achieved.

Finally, if it is agreed that convergence ceases when M_∞ exceeds, by a certain small but nonzero amount, the "lower critical" Mach number (for which the critical speed is first attained in the flow), then, since experimental results and calculations by the hodograph method (Sec. F) suggest that the existence of continuous adiabatic flow ceases under similar conditions, and since in any case convergence could not be expected to continue after such continuous flow had broken down, it is reasonable, until evidence to the contrary is found, to suppose that the two events occur simultaneously. However, the methods could hardly be used for values of M_∞ just before that at which a shock becomes necessary, as convergence would doubtless be too slow.

A successive approximation method for small disturbances to a uniform stream. Another method for treating flows which are subsonic at least at large distances was described by Görtler [*10*] and by Hantzsche and Wendt [*11*]. The limitations on convergence appear to be similar to those just stated, but the method is nevertheless greatly superior to those described above in problems where it is suitable, namely, for evaluating the flow past obstacles whose breadth is considerably smaller than their length, but not so much smaller that the Prandtl-Glauert small perturbation theory can be applied. This latter theory, which neglects the squares of the perturbations, is treated as a first approximation to the flow, and higher approximations are deduced by neglecting only the cubes of the perturbations, and so on.

Putting $\phi = q_\infty(x + \varphi)$, so that $q_\infty\varphi$ represents the potential of disturbances to the main stream, Eq. 2-1 may be approximated, neglecting the cubes of disturbances, as

$$\nabla^2\varphi - M_\infty^2\varphi_{xx} = (\gamma - 1)M_\infty^2\varphi_x\nabla^2\varphi + 2M_\infty^2\nabla\varphi \cdot \nabla\varphi_x \qquad (2\text{-}14)$$

(Eq. 2-14 would be exact if φ_x, in its two occurrenees on the right, were replaced by $\varphi_x + \frac{1}{2}(\nabla\varphi)^2$ and $\varphi_x + \frac{1}{4}(\nabla\varphi)^2$ respectively.)

If φ_1 is the first approximation to the disturbance potential φ, given by the small perturbation theory, in other words by approximating the right-hand side of Eq. 2-14 by zero, under the boundary condition of zero normal velocity at solid surfaces, then the second approximation φ_2 is obtained by solving Eq. 2-14 for φ, under the same boundary condition, with φ replaced throughout the right-hand side by φ_1. For the third approximation φ_3 the same process is gone through, but with the right-hand side of Eq. 2-14 rendered exact as indicated in the sentence following it, and with φ replaced therein by φ_2. This approximation is as far as any author has pursued the calculations, but it is clear how they would be continued. At each stage an equation of Poisson's type has to be solved, in the region outside the body stretched in the direction of motion by a factor $(1 - M_\infty^2)^{-\frac{1}{2}}$. The analysis is greatly simplified by Kaplan's discovery [*53*] of general expressions for particular integrals of the equations for φ_2 and φ_3, in terms of the φ's already found.

Hantzsche and Wendt [*11*], by calculating φ_2 for various plane flows, were able to give a general picture of the influence of thickness, camber, and angle of attack on the compressibility effect. Actually (unlike

Görtler [*10*]), they work with the stream function ψ rather than the potential ϕ, and this has definite advantages in the present method, in which it is much easier to apply a boundary condition in the form $\psi = 0$ than one in the form $\partial\phi/\partial n = 0$.

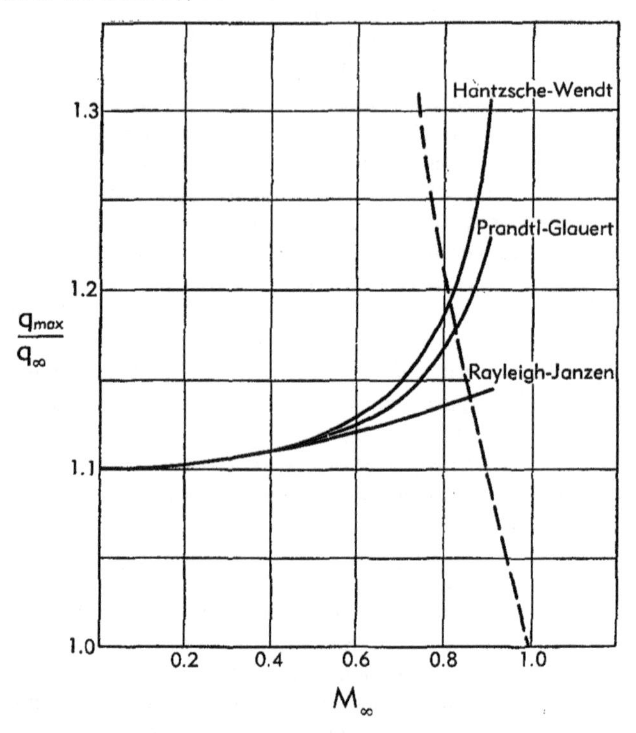

Fig. E,2b. Maximum surface velocity in symmetrical flow at Mach number M_∞ past an ellipse of thickness ratio 0.1.

For example, they calculate the maximum velocity $q_{\max}$ in the symmetrical flow past an ellipse of thickness ratio ϵ as

$$q_{\max} = q_\infty \left[1 + \frac{\epsilon}{\sqrt{1 - M_\infty^2}} + \frac{\epsilon^2 M_\infty^2 (1 - \frac{3}{8} M_\infty^2)}{2(1 - M_\infty^2)^2} + O(\epsilon^3) \right] \quad (2\text{-}15)$$

It is likely that, for a given value of M_∞ less than 1, the series (2-15) converges when the thickness ratio ϵ is small enough for a continuous flow to be possible. For M_∞ near 1 the order of the second and third terms in the series indicates that the limiting value of ϵ is likely to be of order $(1 - M_\infty^2)^{\frac{3}{2}}$; and this is confirmed by the transonic similarity theory (D,33). The ratio of maximum surface velocity to main stream speed, $q_{\max}/q_\infty$, given by Eq. 2-15, is graphed against Mach number for $\epsilon = 0.1$ in Fig. E,2b. It is compared with the value $1 + \epsilon/\sqrt{1 - M_\infty^2}$ given by the Prandtl-Glauert small perturbation theory, and with the value given by

Kaplan's calculation [6] of the second Janzen-Rayleigh approximation. The broken line represents critical conditions, where the local Mach number is 1; thus it is the graph of q^*/q_∞ against M_∞ as given by Bernoulli's equation.

It is fairly clear that the second Hantzsche-Wendt approximation is superior to the second Janzen-Rayleigh approximation. For the first approximation is certainly better, since it yields Glauert's equation rather than Laplace's. And the procedure to the second approximation is more accurate in two respects: (1) in the equation for φ_2 in terms of φ_1 a term $-M_\infty^2\varphi_{2xx}$ is included where the Janzen-Rayleigh method would have $-M_\infty^2\varphi_{1xx}$; (2) the variation of the velocity of sound is taken into account by including the term $(\gamma - 1)M_\infty^2\varphi_{1x}\nabla^2\varphi_1$. These are likely to be far more significant than the third point of difference, which works the other way: (3) the term $2M^2\nabla\varphi_1 \cdot \nabla\varphi_{1x}$ is less accurate than the term $2M^2\nabla\varphi_1 \cdot \nabla[\varphi_{1x} + \frac{1}{4}(\nabla\varphi_1)^2]$ occurring in the equation for the Janzen-Rayleigh second approximation as rewritten in terms of φ.

Probably, for bodies whose thickness ratio is of the same order as 1/10, which is that of the ellipse of Fig. E,2b, the second Hantzsche-Wendt approximation is reasonably accurate in neglecting its cube for all subcritical Mach numbers; e.g. for $M_\infty = 0.81$ the terms inside the square brackets in Eq. 2-15 are $1 + 0.171 + 0.020$. Rapid convergence is indicated, to a value close to the critical value 1.198. This indicates not only that the lower critical Mach number for symmetrical flow past an ellipse of thickness ratio 1/10 is 0.81, but also that all calculations of φ_2 for shapes of practical importance should be useful.

Hantzsche and Wendt [11], by calculating φ_2 for a circular arc airfoil of zero thickness, were able to estimate the influence of camber on the compressibility effect. Its influence on the maximum velocity achieved is similar to that of thickness, increasing this quantity above the Prandtl-Glauert value. However, it has no influence on the lift at a given angle of attack, which is given by Glauert's law even to this second approximation if only the camber and not the thickness of an airfoil is taken into consideration.

To estimate the influence of thickness on lift, φ_2 was calculated for a symmetrical Joukowsky airfoil. The method of calculation makes it necessary to assume that the airfoil, when stretched in the direction of motion by a factor $(1 - M_\infty^2)^{-\frac{1}{2}}$, is still a Joukowsky airfoil, but this is true to a good approximation for fairly thin airfoils. For thickness ratio ϵ, the ratio of the lift coefficient to its value for zero Mach number is

$$\frac{C_L}{(C_L)_{M_\infty=0}} = \frac{1}{\sqrt{1 - M_\infty^2}} + \frac{4\epsilon}{3\sqrt{3}} \frac{(1 - M_\infty^2) - (1 - M_\infty^2)^{\frac{3}{2}} + \frac{1}{4}(\gamma + 1)M_\infty^4}{(1 - M_\infty^2)^2} \quad (2\text{-}16)$$

The actual value of $(C_L)_{M_\infty=0}$, with the theoretical value of the circulation sufficient to give a finite velocity at the trailing edge without any boundary layer, is $(1 + 4\epsilon/3\sqrt{3})2\pi \sin \alpha$. Thus Eq. 2-16 states that, while the basic flat plate lift coefficient $2\pi \sin \alpha$ is increased by the factor $(1 - M_\infty^2)^{-\frac{1}{2}}$ as a result of compressibility, the *addition to the lift coefficient due to thickness* is increased by the factor

$$\frac{1 - M_\infty^2 + \frac{1}{4}(\gamma + 1)M_\infty^4}{(1 - M_\infty^2)^2} \tag{2-17}$$

Kaplan [*12*] gives reasons for believing that this latter result is true, to within the Hantzsche-Wendt approximation, for all airfoils.

However, the concept of an "additional lift due to thickness" is one of the conclusions of potential theory bearing least relation to reality. The boundary layer thickness increases so much with airfoil thickness (owing to the larger positive pressure gradient over the rear portion of the surface) that the reduction due to this in the cast-off vorticity necessary to achieve smooth flow at the trailing edge easily balances the increase due to thickness derived from potential theory. For some sections the lift curve slope is less than 6; only for a few modern "low drag" sections does it exceed 2π. As for wind tunnel observations on the increase of lift curve slope with Mach number (which, it may be remarked, continues noticeably beyond the lower critical value), the Prandtl-Glauert curve is usually within the limits of experimental error, and there appears to be no conclusive evidence that the rate of increase which it gives is ever exceeded. It should be remembered, however, that, if ever effective methods of boundary layer control are used, then additional lifts as discussed above should be observed.

Görtler [*10*] applied the method under discussion to flow along a wavy wall, and by simplifying the boundary condition was able to proceed to the third approximation φ_3. The boundary condition adopted for the disturbance potential φ was

$$\varphi = K \cos \frac{2\pi x}{l} \qquad \text{on} \quad y = 0 \tag{2-18}$$

which indicates approximately a solid wavy wall with wavelength l on $y = 0$; the flow is to be determined for $y > 0$, and since the velocity is q_∞ at infinity, φ has zero gradient at large distances from the wall. The precise shape of the wall remains to be determined. For $M_\infty = 0.9$ and $K = 0.035l$ the first three approximations were calculated. They appear to be converging, and the flow indicated has a local supersonic region attached to each place where the wall is convex to the stream. In Fig. E,2c a section of the flow, comprising a single wavelength, is illustrated by a plot of isobars: the figures on these represent the ratio of the fluid speed q to the critical speed q^*, a ratio which exceeds 1 only if the flow is locally supersonic. The wall shape is not sinusoidal; the troughs are considerably deeper and very much "sharper" than the peaks.

Görtler points out that, since in the local supersonic regions the streamlines must diverge with increasing velocity, whereas the approximation φ_1 is essentially of subsonic type with convergence of streamlines in accelerating flow, it is clear that, when the present method is applied to transonic problems, the new terms in the second approximation φ_2 must be of the same order of magnitude as the old terms (i.e. those in φ_1). On

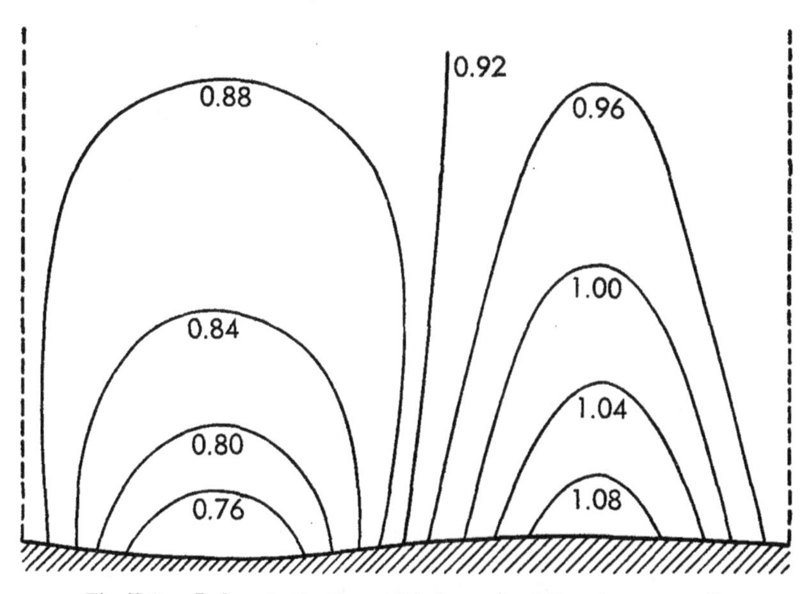

Fig. E,2c. Isobars in the flow at Mach number 0.9 past a wavy wall.

the other hand it might be hoped that φ_3 would represent only a small correction to φ_2, and this is observed in the example computed.

In a footnote, Görtler remarks that the solution was also attempted with an alternative boundary condition, analogous to Eq. 2-18 but prescribing $\partial\varphi/\partial y$ on $y = 0$ rather than φ. However, there was no indication of convergence for values of the parameters such that local supersonic regions were present. The physical significance of this may be that the corresponding wall shape would be much more nearly sinusoidal, and that for such shapes (as against that of Fig. E,2c) the breakdown of continuous flow occurs for a lower value of M_∞.

Expansion in powers of a thickness parameter. Before passing from the Hantzsche-Wendt method it should be remarked that a method identical in its aims, and indeed appearing at first sight to constitute the carrying of the method of Hantzsche and Wendt to its logical conclusion, had been tried by them and also by Kaplan [*12*] but subsequently discarded. It is directly analogous to the Janzen-Rayleigh method in that the disturbance potential φ is supposed expanded in powers of some parameter, say ϵ, defining the magnitude of the disturbance. For flow past an obstacle it would be convenient to take ϵ as the thickness ratio of the obstacle. The flow past a whole class of obstacles, obtained by shrinking the given obstacle by a uniform factor, but only in the direction perpendicular to the flow, is considered. Then the disturbance potential φ depends on ϵ, which varies from 0 to ∞ for this class of flows, and an expansion

$$\varphi = \epsilon\varphi^{(1)} + \epsilon^2\varphi^{(2)} + \cdots \tag{2-19}$$

can reasonably be attempted, where $\varphi^{(1)}, \varphi^{(2)}, \ldots$ are functions of x and y depending on M_∞. They are determined by substituting Eq. 2-19 in the equation for the disturbance potential *and in the boundary conditions*, and equating coefficients of powers of ϵ. It is this apparently logical conception, of approximating to the boundary condition as well as to the equation of motion, which causes the method to fail in important cases. If the equation of the obstacle, supposed two-dimensional, is $y = \epsilon Y(x)$, then the boundary condition is

$$\frac{v}{u} = \frac{\partial\varphi/\partial y}{1 + \partial\varphi/\partial x} = \epsilon Y'(x) \quad \text{on} \quad y = \epsilon Y(x) \tag{2-20}$$

To expand this completely in powers of ϵ, it is necessary to replace the values of any function, say $\partial\varphi/\partial y$, on $y = \epsilon Y(x)$, in terms of the values of it and its derivatives on $y = 0$, namely, by

$$\left[\frac{\partial\varphi}{\partial y} + \epsilon Y(x)\frac{\partial^2\varphi}{\partial y^2} + \frac{1}{2}\epsilon^2 Y^2(x)\frac{\partial^3\varphi}{\partial y^3} + \cdots\right]_{y=0} \tag{2-21}$$

As a result the boundary conditions on $\varphi^{(1)}, \varphi^{(2)}, \ldots$, obtained by equating coefficients of $\epsilon, \epsilon^2, \ldots$ in condition (2-20), become

$$\frac{\partial\varphi^{(1)}}{\partial y} = Y'(x), \qquad \frac{\partial\varphi^{(2)}}{\partial y} = Y'(x)\frac{\partial\varphi^{(1)}}{\partial y} - Y(x)\frac{\partial^2\varphi^{(1)}}{\partial y^2}, \cdots \quad \text{on } y = 0 \tag{2-22}$$

The equations to be solved for $\varphi^{(1)}, \varphi^{(2)}, \ldots$ are almost identical with those in the Hantzsche-Wendt method, but the boundary conditions are applied on a straight strip in the direction of flow, coextensive with the obstacle.

Such boundary conditions are in general easier to apply. However, in the important case when the obstacle, though thin, has a blunt nose, the flow over which has a stagnation point, the method fails. It does this even for $M_\infty = 0$, i.e. when used as a method for obtaining the incompressible potential flow around airfoils. The first approximation $\epsilon\varphi^{(1)}$ then represents the first order airfoil theory, already given, for camber lines only, by Glauert [*13*, p. 87]. The flow is reasonably approximated (for small ϵ) on this theory except near the leading edge. Here there is a square root singularity of $\varphi^{(1)}$, and its gradient (representing the velocity) is infinite. However, for many purposes the errors in this neighborhood do not matter, and the over-all picture, which is known by a comparison with Theodorsen's more exact airfoil theory [*14*] to be a correct first approximation, is all that is required. But when $\varphi^{(2)}$ is calculated, using the second of conditions (2-22), it seems likely from the form of this condition that a more severe singularity will appear in $\varphi^{(2)}$ than in $\varphi^{(1)}$. Actually, however, the presence of this more severe singularity can be avoided in $\varphi^{(2)}$ by fixing the origin at the leading edge, so that $Y(x) = 0$ there; but it is unavoidable in $\varphi^{(3)}$. It may be presumed that the order of the singularity grows with successive terms. Hence, at least sufficiently near the leading edge, the process cannot converge. Therefore, since the process is not tending to a limit satisfying all the prescribed conditions, there is no reason why a later term should be more accurate than an earlier term.

Hence the method is not applicable, at any rate without modification, to bodies with blunt noses. For these it would therefore appear to be necessary to use the Hantzsche-Wendt method sketched above, in which the boundary condition is applied exactly on the surface of the body. The only successful investigation on the lines just described has been Kaplan's on the flow past a smooth hump in a wall [*52*], for which the difficulties mentioned do not arise.

Lighthill [*56*] has shown (using the method described at the end of Art. 1) that the logical process of expansion in powers of ϵ gives, when $M_\infty = 0$, and when terms in ϵ and ϵ^2 only are included, an approximation valid uniformly, even near the leading edge, provided the deduced flow field is shifted downstream through a distance equal to half the radius of curvature of the profile at the leading edge. It is possible that a similar result may be true for the case of general subsonic Mach number, which however is not investigated in the paper cited, and the point is still unsettled. If this were so, it would produce a method for calculating second order compressibility effects, due to the thickness, camber, and incidence of airfoils, simpler than the Hantzsche-Wendt process. But, for still higher order terms, the method of expansion (2-19) is certainly impossible, since Hantzsche showed [*15*] that for flow past an elliptic cylinder terms in $\epsilon^4 \ln \epsilon$ appear.

Comparison with the Kármán-Tsien method. Considering the results of all the methods so far described as a whole, it may be claimed that a fair picture of shock-free subsonic flow has been achieved. Enough particular solutions have been found to make possible an adequate guess in any given problem of how the compressibility effect predicted by potential theory will depart from that given by the small perturbation theory.

However, the material has in fact been little used in practical high speed flow work, because it is somewhat extensive. It is much easier to take a single formula and apply it under all conditions, and such a formula was provided by Tsien [*16*], who worked on an idea of von Kármán's.

The derivation of the Kármán-Tsien approximation is described in detail in F, 5 but its importance (due to its intrinsic interest as well as to the fact that it is by far the most widely used formula relating to two-dimensional subsonic flows) is so great that a comparison of its results with those of the other methods (described above) is here made.

The approximation made to the equations of motion by von Kármán was equivalent to taking full account of the finiteness of the velocity of sound and of the convection of sound by the local motion of the fluid, but adopting a quite different mode of variation of the speed of sound with the density (and hence with the local fluid speed) from the true one. Thus the velocity of sound is taken proportional to a negative power of the density, $a/a_\infty = (\rho/\rho_\infty)^{-1}$ instead of to a small positive power, $a/a_\infty = (\rho/\rho_\infty)^{\frac{1}{2}(\gamma-1)}$. The equation of motion (2-1) is therefore modified by the coefficient a^2 of $\nabla^2\phi$ being replaced by $a_\infty^2 - (q_\infty^2 - q^2)$; in short, γ is therein replaced by -1.

It is at once clear that exact solutions of the resulting equation for the velocity field (which can be obtained fairly easily, although, as will be seen, the obstacle shape cannot be specified in advance) must give the first approximation to the effect of small nonzero Mach number quite correctly. In other words, they are at least the equal of the second Janzen-Rayleigh approximation. For it was seen above that this was influenced only by the terms due to the convection of sound on the right of Eq. 2-1, and not at all by the variation of the speed of sound with q. In fact Eq. 2-5 for ϕ_1 is unaltered by replacing γ by -1.

But actually the exact solution with γ replaced by -1 is probably always better than the second Janzen-Rayleigh approximation, because more accuracy is gained by accurately admitting the convection of sound than is lost by allowing the speed of sound to vary incorrectly. That this is so for very thick obstacles is indicated by Eq. 2-8 for the maximum speed over a circular cylinder, which is closer to its exact value with γ replaced by -1 than with all the terms in M_∞^4, M_∞^6, . . . omitted. For very thin obstacles, on the other hand, the equation of von Kármán becomes identical with that of Prandtl and Glauert (which again is independent of γ), and the conclusions are again superior to those of the second Janzen-Rayleigh approximation (which replaces $(1 - M_\infty^2)^{-\frac{1}{2}}$ by $1 + \frac{1}{2}M_\infty^2$).

For obstacles which are only fairly thin the equation probably gives results intermediate in accuracy between the Prandtl-Glauert small perturbation theory and the second Hantzsche-Wendt approximation.

For on this latter approximation, the disturbance potential φ is replaced by φ_1 in the right-hand side of Eq. 2-14 and hence also $\nabla^2\varphi$ may be replaced by $M_\infty^2\varphi_{1xx}$. The relative importance of the two terms is therefore probably as $(\gamma - 1)M_\infty^4 : 2M_\infty^2$, and so the change of γ from 1.4 to -1 probably does not actually reverse the effect of φ_2 for any $M_\infty < 1$, and for smaller M_∞ may give a good approximation to φ_2. Roughly then, the replacement of γ by -1 deals accurately with the compressibility effect on very thin obstacles, but approximates *as for small Mach number* to the additional compressibility effect due to the finiteness of the disturbances.

The solution of Eq. 2-1 with γ replaced by -1, which is here discussed only for plane flows without circulation, is as follows (F, 5): If the velocity vector in incompressible potential flow about an obstacle is $(u_i,\ v_i)$ at x_i, y_i—here the subscript i stands for "incompressible"—then the velocity vector in "compressible flow with γ replaced by -1" about a certain obstacle of modified shape, with the same speed in the undisturbed stream, is

$$(u,\ v) = \frac{(u_i,\ v_i)}{1 + \frac{1}{2}[1 - (1 - M_\infty^2)^{-\frac{1}{2}}]\left(\frac{u_i^2 + v_i^2}{q_\infty^2} - 1\right)} \tag{2-23}$$

at the point x, y given by

$$x + iy = x_i + iy_i - \frac{M_\infty^2}{[1 + (1 - M_\infty^2)^{\frac{1}{2}}]^2}\int\left(\frac{u_i + iv_i}{q_\infty}\right)^2 (dx_i - idy_i) \tag{2-24}$$

Eq. 2-24 represents a mapping of the x_i, y_i plane onto the x, y plane. Streamlines correspond under this mapping, and therefore the flow in the x, y plane is the flow about an obstacle obtained by carrying out the mapping on the area outside the original obstacle.

It is seen that the formulation of the solution is brief. Yet the solution is certainly just as accurate as the second Janzen-Rayleigh approximation (even for thick obstacles, and more so for thin ones), for which the analytic solution for a general shape of obstacle [7] takes a far more complicated form. This is essentially because a deformation of the body shape (and, with it, of the whole physical plane), as well as of the velocity variations, has been permitted.

However, the exact solution just stated is not the one advocated by Tsien and used widely in aeronautics. The Tsien approximation is to adopt the formula (2-23) defining the compressibility correction to the velocity vector, but to ignore the deformation of the obstacle and the flow plane by the mapping (2-24). To gauge the significance of this, observe that, although the obstacle is shrunk by the mapping (2-24), a uniform shrinkage (i.e. a mere change of scale) would not alter the velocity distribution on potential theory. It is the nonuniformity of shrinkage in the mapping that causes errors in neglecting it.

Now a small elementary arc of a streamline is shrunk by a factor

$$1 - \frac{M_\infty^2}{[1 + (1 - M_\infty^2)^{\frac{1}{2}}]^2} \frac{(u_i^2 + v_i^2)}{q_\infty^2} \tag{2-25}$$

(but unaltered in direction) in the mapping (2-24). Thus for Mach numbers as high as 0.8, for which the fraction involving the Mach number in Eq. 2-25 is $\frac{1}{4}$, the shrinkage near the point of maximum velocity is considerable, while there is none at the stagnation point. (On the other hand, small elementary arcs normal to the streamlines are *expanded* in the mapping by a factor identical with Eq. 2-25 but with the sign of the second term changed; this is less relevant, however, from the practical point of view, since one is mainly interested in conditions on the surface of the obstacle, which constitutes a streamline.) The obstacle as modified by the mapping therefore has a greater thickness ratio.

The correctness to order M_∞^2, for bodies of arbitrary thickness, of the complete solution given by the "velocity correction formula" (2-23) and the mapping (2-24), is lost by dropping the latter, since the mapping changes the flow field by amounts of order M_∞^2. However, the Tsien approximation was put forward only for dealing with the case of fairly thin obstacles, which constitutes the major field of practical application of potential flow theory.

But even here, and even if we temporarily ignore the difficulties associated with the perturbations not being small for practical obstacles with leading edge stagnation points, there is a serious loss of accuracy due to dropping the mapping (2-24). For if, as before, ϵ denotes the thickness ratio of the obstacle and hence the order of magnitude of the disturbances, the expression (2-25) indicates that differences in relative shrinkage can be of order ϵM_∞^2. Since the main part of the disturbance potential, derived from the small perturbation theory, is of order ϵ, it follows that changes in it due to the differential shrinkage are of order $\epsilon^2 M_\infty^2$. But the difference between the Prandtl-Glauert and second Hantzsche-Wendt approximations is itself of order $\epsilon^2 M_\infty^2$. Hence the Kármán-Tsien approximation retains no advantage over the Prandtl-Glauert approximations as far as strict mathematical orders of magnitude are concerned, the error in both being of order $\epsilon^2 M_\infty^2$ in the disturbance velocities (while the full solution of Eq. 2-23 with Eq. 2-24 contains the correct term in $\epsilon^2 M_\infty^2$, the error being of order $\epsilon^2 M_\infty^4$).

However, this discussion of mathematical orders of magnitude is misleading if it leads one to suppose that the velocity field is given just as accurately by the small perturbation theory as by the Kármán-Tsien formula. In practically all tests of airfoils in high speed wind tunnels the surface pressure distribution as deduced from the Kármán-Tsien formula is within the limits of experimental error for subcritical Mach numbers, while that deduced from the Prandtl-Glauert formula is not so for the higher Mach numbers (say $M_\infty \geqq 0.5$).

It is commonly suggested that this is true, in spite of the identity in mathematical order of magnitude of the two errors, for two reasons: (1) the fraction containing M_∞ in Eq. 2-25 remains rather smaller for subcritical Mach numbers than its order of magnitude "M_∞^2" would suggest, because it is approximately $\frac{1}{4}M_\infty^2$ for small M_∞ and is only 0.25 for $M_\infty = 0.8$, which is about the highest lower critical Mach number occurring in practice; (2) the errors due to allowing the velocity of sound a to vary in an incorrect manner, and due to neglecting the mapping (2-24), are of opposite sign and partly cancel out, although probably the latter is larger at lower Mach numbers and the former at higher Mach numbers. There is some evidence that the canceling suggested in the latter explanation does occur on the suction side of the airfoil, which is probably the critical region. For one may argue that the equations solved are the equations of continuity and irrotationality (2-12), with an incorrect relationship between density and fluid speed replacing Eq. 2-13. This incorrect relationship (obtained by replacing γ by -1) consistently overestimates the density for given q/q_∞ other than unity, by a factor increasing approximately like $1 + \frac{1}{8}(\gamma + 1)(q^2 - q_\infty^2)^2/a_\infty^4$, as q departs from q_∞ on either side. But on the suction side the error due to this might be partly canceled by the error due to ignoring the mapping (2-24), owing to the fact that the area of a small rectangle in the network of streamlines and equipotentials, on a scale such that it is unaltered for $q = q_\infty$, is changed in the mapping by a factor less than 1 by a certain multiple of $q_\infty^2(q^2 - q_\infty^2)/a_\infty^4$. These two factors, when multiplied to give the change in mass in such a small rectangle, tend to cancel in part, doing so completely for $q = 1.19q_\infty$ at small Mach numbers and for $q = 1.35q_\infty$ at $M_\infty = 0.8$. But the importance of this point can at present only be described as conjectural.

Actually, it is with measurements of pressure, not of velocity, that the theoretical results are compared. In using the expression (2-23) for the velocity field to deduce the density field and hence the pressure field, Tsien [*16*] recommends that the pressure-density relation should be taken as that obtained by integrating the assumed relation between a^2 and ρ, and taking $p = p_\infty$ in the undisturbed stream. This gives a linear relation $p/p_\infty = 1 + \gamma(1 - \rho_\infty/\rho)$ between pressure p and specific volume $1/\rho$. The relation between the pressure coefficient $C_p = (p - p_\infty)/\frac{1}{2}\rho_\infty q_\infty^2$ and the fluid speed is then not the exact relation for adiabatic flow, namely,

$$C_p = \frac{[1 + \frac{1}{2}(\gamma - 1)M_\infty^2(1 - q^2/q_\infty^2)]^{\gamma/(\gamma-1)} - 1}{\frac{1}{2}\gamma M_\infty^2} \tag{2-26}$$

but is what is obtained by replacing γ by -1 in Eq. 2-26. This latter relation, with Eq. 2-23, gives the following very simple relation between the pressure coefficient in two-dimensional compressible flow, C_p, and that in the incompressible flow past the same obstacle, C_{pi}:

$$C_p = \frac{C_{pi}}{(1 - M_\infty^2)^{\frac{1}{2}} + \frac{1}{2}[1 - (1 - M_\infty^2)^{\frac{1}{2}}]C_{pi}} \tag{2-27}$$

This is the formula mentioned above as so extensively used in aeronautics, and later as fitting wind tunnel measurements in subcritical conditions to within the experimental error.

The reader may perhaps imagine at this stage that the further error due to replacing γ by -1 in Eq. 2-26 is of importance in the discussion of the total deviation of the Kármán-Tsien pressure field from the exact field which would occur in adiabatic flow. For the Kármán-Tsien theory is essentially one relating to the velocity and density fields; it establishes by kinematic considerations the irrotational flow in which the density and fluid speed are connected in a certain way. Thus it would seem inadvisable, from a general point of view, to add to errors in Eq. 2-23 by using an inaccurate form of Eq. 2-26; but on the other hand a surprising accuracy in the final result might seem explicable by supposing that the previous errors are partly canceled by the error in using the said form of Eq. 2-26.

However, there is relatively little substance in these suspicions, for the simple reason that the error in replacing γ by -1 in Eq. 2-26 is always rather small. When Eq. 2-26 is expanded in powers of the disturbances, the coefficients of the first and second powers are independent of γ and the error in changing γ from 1.4 to -1 in the third term is $\frac{1}{10}M_\infty^4(1 - q^2/q_\infty^2)^3$. For small M_∞ or ϵ this term of order $M_\infty^4\epsilon^3$ is smaller than the errors of order $M_\infty^4\epsilon^2$ and $M_\infty^2\epsilon^2$ which were mentioned above. And indeed, even in a rather extreme case when $q^2/q_\infty^2 = 2.5$ and $M_\infty = 0.8$, expression (2-26) is -1.174 for $\gamma = 1.4$ but -1.111 for $\gamma = -1$, a difference of only 5.4 per cent.

But the formula relating the lower critical Mach number M_{cr} to the negative pressure coefficient $(C_p)_{min}$ with largest numerical magnitude on the surface, namely

$$(C_p)_{min} = \frac{2}{\gamma M_{cr}^2}\left[\left(\frac{2 + (\gamma - 1)M_{cr}^2}{\gamma + 1}\right)^{\gamma/(\gamma-1)} - 1\right] \tag{2-28}$$

cannot be simplified in such a way. Tsien recommends that Eq. 2-28 should be used in its exact form in conjunction with the experimentally verified formula (2-27). This gives a relation between the minimum pressure coefficient $(C_{p_i})_{min}$ in the *low speed* potential flow past an obstacle and the lower critical Mach number M_{cr} for flow past the obstacle, which is shown graphically in Fig. E,2d. This curve is extensively used in airfoil design, since (as follows from statements already made above) it is almost always in agreement with wind tunnel experiments to within the orders of accuracy obtainable.

To sum up this article, considerable information on the development of flow patterns with Mach number until just before shock waves appear is obtainable; but the methods described are useless for investigating its further development. Indeed, as will be seen from other parts of this work, our theoretical understanding of this further development, at least up to where the Mach number of the main stream reaches unity, is very restricted indeed.

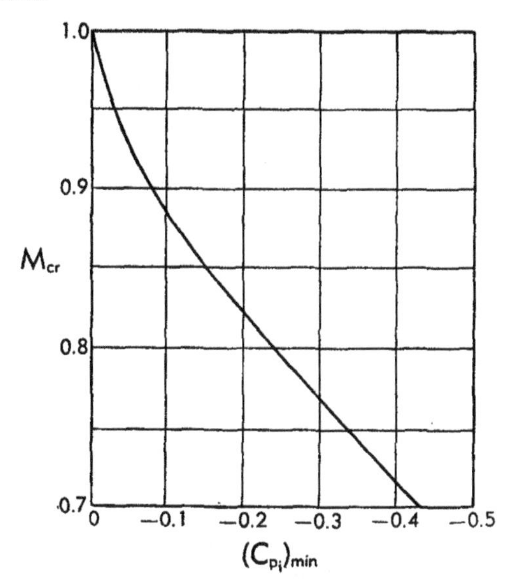

Fig. E,2d. Tsien's curve of lower critical Mach number against minimum pressure coefficient for incompressible flow.

E,3. Supersonic Two-dimensional Airfoil Theory: Surface Pressures. The theory of higher approximations than the acoustic "small perturbation" theory is more extensive in the supersonic than in the subsonic regime. For this reason it is here divided into the part relating to two-dimensional flow (Art. 3, 4, and 5), for which a remarkably detailed picture can be obtained, and the part relating to three-dimensional (and especially axisymmetrical) flows, of which our understanding is less complete. In Art. 3, 4, and 5, then, we describe those improvements which have been made to the picture given by Ackeret's small perturbation theory of the steady two-dimensional flow of a uniform supersonic stream past an airfoil, without taking account of the diffusion of shear and heat in the boundary layer.

The reason why the nonlinearity associated with the variable speed of sound, and with the convection of sound waves by the local motion of the fluid, is more easily dealt with in supersonic than in subsonic flow, is that every point in the flow has only a limited domain of influence (the downstream characteristic conoid). This already makes the small perturbation theory simpler. (For example, numerous calculations on exact lifting surface theory have been made in the supersonic regime (Sec. D), but these are almost prohibitively lengthy for the simplest planforms in the subsonic regime.) It is not surprising that this greater simplicity should penetrate also to the higher approximations.

Nevertheless, there is a third source of nonlinearity in supersonic flow, as mentioned in Art. 1, namely, the shock (the need for whose presence springs from a combination of the other two sources). As will be seen below, an attempt to estimate the effect of those properties of the shock which pressure discontinuities in the acoustic theory do not possess, especially the partial reflection of waves which overtake it and the gain in specific entropy of the fluid passing through it, adds considerably to the complexity of the theory, although here it has been carried fairly far for two-dimensional flow past airfoils. However, it will also be found that the properties of the shock just mentioned can be neglected while retaining a much better approximation to the flow field than is given by the small perturbation theory.

It should be remembered that in supersonic flow the use of a two-dimensional theory of flow past an isolated section may be of value not merely as an approximation, but also for an accurate study of the flow past wings, whether in a wind tunnel or in free flight. For in a two-dimensional wind tunnel test there will be no tunnel interference provided that the shocks from the leading edge have reflections at the walls which intersect behind the airfoil; for then the flow within the curvilinear quadrilateral formed by the shocks, and in particular the surface pressure distribution, will be given correctly by a two-dimensional theory in which the incident stream is supposed uniform and infinitely wide. Or again, in the uniform supersonic flight of a wing of uniform cross section and rectangular planform, a portion of the flow, namely, that which is outside the domains of influence of the endpoints of the leading edge, will be two-dimensional.

During the whole of this article and of Art. 4 we confine ourselves to the study of airfoil sections which are sharply pointed at both their leading and trailing edges. Further, we suppose that the angle of attack and the Mach number M_∞ of the incident stream are so limited that any deflections of the stream through angles concave to the flow, which have to be made at the leading and trailing edges, can be achieved with the aid of attached shocks with supersonic flow on both sides of them. The condition necessary for this is fully studied in III,E,2; for given Mach number, the deflection must not be too large; for given deflection, the Mach number must not be too small. If we make the restriction that the shocks are to be attached, we might just as well require also that the flow behind them be supersonic, since this greatly assists the theory, while (for all Mach numbers) it only limits further the range of possible deflections by less than half a degree (the maximum percentage change in the greatest permitted deflection being 8 per cent, attained in the limit $M_\infty \to 1$, and the maximum total change $0°25'$, attained for $M_\infty = 1.4$). Only in Art. 5 is the effect of a blunt leading edge discussed.

It follows from the assumption just introduced that there is no mutual influence between the flow over the upper and lower surfaces of the airfoil. For the incident stream is uniform ahead of the sharp leading edge, where it divides into two streams, above and below the airfoil, which cannot interact until they reunite behind the trailing edge. Since the flow is assumed supersonic even behind the trailing edge, this interaction by union has a domain of influence entirely downstream of the trailing edge, and it cannot therefore affect the surface pressures. Hence for the purpose of calculating these, which forms the subject of the present article, the flow at and beyond the trailing edge need not be considered. Only in Art. 4 is the complete flow pattern discussed. (But it must be mentioned that the principle just established is the point at which the real flow is worst approximated by a theory neglecting the boundary layer. The conditions at the trailing edge do influence the pressures upstream of it, by propagation up the boundary layer, particularly when it is laminar. In making predictions from the adiabatic theory it is therefore important to make allowance for this, as discussed in IV,B.)

Prandtl-Meyer flow. It is seen, then, that on adiabatic theory the problem of determining the flow over, say, the upper surface is equivalent to that of determining how a uniform supersonic stream parallel to a plane wall is disturbed at and behind a corner (corresponding to the leading edge), where the wall bends abruptly and then curves in an arbitrary two-dimensional manner. This problem was already considered by Prandtl [*17*] in 1907, in connection with the flow of steam in Laval nozzles. He points out that the flow round a corner concave to the flow can be achieved by means of a plane oblique shock (provided that the required deflection is less than a quantity dependent on the Mach number to which reference was made above), but that some other mechanism is necessary to produce flow round a corner convex to the flow. He argues that, since the state of affairs behind the corner cannot influence the flow around it, the boundary conditions define no length scale proper to the problem. Therefore, it is reasonable to seek a solution in which every physical quantity is constant along radii through the corner, and which depends only on the angular polar coordinate with the corner as pole. He states that he has found solutions of this type to the equations of potential flow, with a polytropic equation of state (which includes, for air, the condition of constant entropy), in which streamlines are bent toward the wall while the gas expands. The complete deflection around the corner is achieved within a wedge-shaped region pointing downstream and away from the wall. Beyond this, if the wall remained straight, then the flow would remain uniform (Fig. E,3a).

Though diagrams of such flows were given in Prandtl's paper [*17*], the formulas for the velocity and thermodynamic variables as functions of the polar angle (see G,4) were first published by Th. Meyer in the subsequent year, and the predicted phenomenon (which corresponds very closely with the observed supersonic flow around corners) is therefore called the Prandtl-Meyer expansion.

Prandtl [*17*] observes that, inside the wedge of expansion, the velocity component perpendicular to a radius vector is always equal to the local speed of sound. This is readily interpreted in terms of the dictum that the equation of irrotational motion is at any one point identical with the equation of sound, in axes moving with the local velocity of the fluid, and with the local value for the speed of sound. Each radius vector then

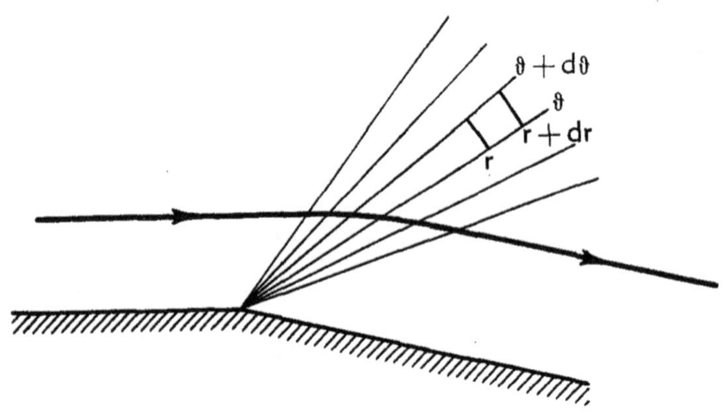

Fig. E,3a. The Prandtl-Meyer expansion.

appears as a plane acoustic wave, which is stationary because its propagation normal to itself at a speed equal to the local value of the speed of sound (which is uniform along it) is exactly balanced by its convection downstream at a velocity whose component normal to the wave is equal and opposite to the velocity of propagation.

This result can be expressed by saying that the angle β between the direction of motion at any point and the local radius vector satisfies $\sin \beta = a/q$ (in terms of the local fluid speed q and speed of sound a). The angle β satisfying this condition is called the local Mach angle; its cosecant is the local Mach number. If θ is the direction of motion (measured in the positive sense from the direction of the undisturbed stream, so that it is negative in the Prandtl-Meyer expansion) and ϑ the polar angle (i.e., in the same sense, the direction of the radius vector), we can write

$$\vartheta = \beta + \theta \tag{3-1}$$

The pressure p, density ρ, local speed of sound a, local fluid speed q, and hence also the Mach angle $\beta = \sin^{-1} (a/q)$, are all functions of one another, by the condition of uniform specific entropy (which applies because the undisturbed stream is uniform) and by Bernoulli's equation. We can find how one of them is related to ϑ (and hence how all of them are, and hence also, by Eq. 3-1, how they are related to the direction of motion θ) by expressing that the mass flowing into a small volume element of space in unit time is equal to that flowing out of it. If the element (Fig. E,3a) is bounded by two radius vectors ϑ and $\vartheta + d\vartheta$ and by two circular arcs with center at the corner and radii r and $r + dr$, and has unit depth perpendicular to the plane of flow, then the mass entering this element (on balance) across the radius vectors is $[d(\rho a)/d\vartheta]d\vartheta dr$, since the velocity normal to a radius vector is a. But, since the velocity and density are uniform along any radius vector, and the velocity component in this direction is $(q^2 - a^2)^{\frac{1}{2}}$ by Pythagoras theorem, the mass flowing out (on balance) across the circular arcs is entirely due to the fact that one is longer than the other by an amount $drd\vartheta$; the said mass is therefore $\rho(q^2 - a^2)^{\frac{1}{2}}drd\vartheta$, and hence

$$\frac{d(\rho a)}{d\vartheta} = \rho(q^2 - a^2)^{\frac{1}{2}} \tag{3-2}$$

With Bernoulli's equation, Eq. 3-2 gives

$$\vartheta = \lambda \tan^{-1}\left[\frac{\lambda a}{(q^2 - a^2)^{\frac{1}{2}}}\right] + \text{const} = \lambda \tan^{-1}(\lambda \tan \beta) + \text{const} \tag{3-3}$$

where $\lambda^2 = (\gamma + 1)/(\gamma - 1)$. Hence by Eq. 3-1 the relation between the direction θ and the magnitude q of the velocity takes the form

$$\theta + \beta - \lambda \tan^{-1}(\lambda \tan \beta) = \text{const} \tag{3-4}$$

$$= \beta_\infty - \lambda \tan^{-1}(\lambda \tan \beta_\infty) \tag{3-5}$$

Here the constant in Eq. 3-3 and 3-4 has been determined by the conditions on the leading radius vector of the expansion, where the flow is still undisturbed, and hence $\theta = 0$.

It was gradually realized, and probably first stated by Ackeret [*18*], that the above arguments, with only slight modifications, are sufficient to determine the deformation of the uniform stream by an arbitrary (two-dimensional) bending of the wall, provided that no shock occurs. For then there is no vorticity or variation of entropy, and a flow is again possible consisting of a single stack of plane acoustic waves, each with all the physical quantities uniform over it. To render each wave stationary, the velocity component normal to it must equal the local speed of sound. The generalization from the Prandtl-Meyer flow around a corner consists only in the fact that the stack of plane waves is not assumed concurrent in a line (the corner, when the wall turns). However the angle ϑ may still be given a meaning, as the angle which the plane wave makes (measured in the positive sense) with the direction of the undisturbed stream. The condition for stationary waves may then again be written in the form of Eq. 3-1. Further, the argument concerning mass flow balance which led to Eq. 3-2 is valid word for word if "plane wave" is read for "radius vector" and if the "corner" (which is taken as the center of the circular arcs bounding the volume element) is replaced by the point (i.e. line, in three dimensions) where the two adjacent plane waves meet.

Thus Eq. 3-3 and 3-5 hold for two-dimensional flow through an arbitrary stack of plane acoustic waves, kept stationary by a balance between their propagation upstream and their convection downstream. This flow

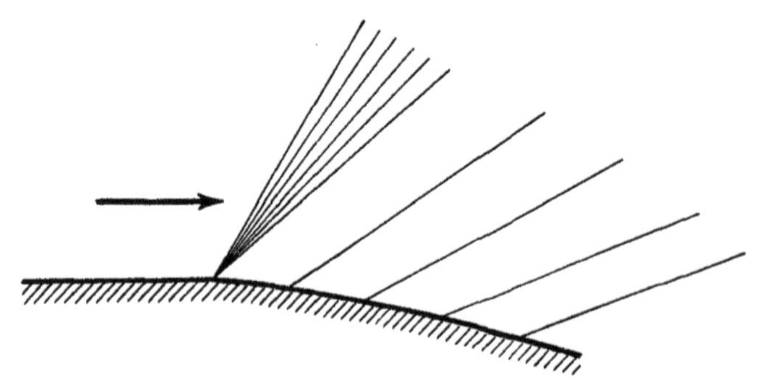

Fig. E,3b. Supersonic flow past a convex wall.

pattern can be made to fit the boundary conditions due to any type of bending of the wall. For these are equivalent to prescribing the direction of motion θ at the wall. Thus, provided that the plane wave through a given point of the wall has the right value of θ, the boundary condition is satisfied. This determines the wave uniquely by Eq. 3-5 and 3-1, which give the Mach angle β on it and the direction of the wave ϑ as a function of θ for a given value of β_∞. Only where the value θ on the wall changes discontinuously (at a corner) is it necessary to have a wedge of expansion where the waves run together.

Such a flow is illustrated in Fig. E,3b. Observed flows in Laval nozzles correspond closely with the predictions of the theory (provided, as is clear from the one-dimensional theory (III,B) of Laval nozzles, that the pressure at the downstream end is low enough to permit the expansion in question without the formation of a normal shock across the nozzle).

Eq. 3-4 (with the constant arbitrary) actually represents the general form of the equations of the characteristics of one of the two systems in any steady two-dimensional supersonic potential flow at constant entropy. (The equations of the characteristics of the other system are the same with $-\theta$ for θ.) Thus the plane waves (or at least the lines in which they cut the plane of the paper in Fig. E,3b) are characteristics. They are also often called Mach lines, since they make the Mach angle β

(whose significance was first realized by E. Mach) with the local direction of flow. (In the more general case of rotational flows this distinguishes them from the third system of characteristics, the streamlines.)

The general type of flow pattern which has here been revealed (and illustrated in Fig. E,3b) is called a "simple wave," because there is one system of characteristics along which the values of physical quantities are propagated unchanged. The properties of simple waves are derived in a different manner, from the general theory of characteristics, in G,6. There, too, a very rapid geometrical method of plotting them, based on Eq. 3-1 and 3-5, is given.

There is a very strong analogy between the simple wave defined above for steady two-dimensional flows, and the simple wave (or "progressive wave") occurring in the theory of nonsteady one-dimensional gas dynamics (III,C). Indeed Fig. E,3b would make an excellent illustration of the latter theory (qualitatively, of course, not quantitatively), provided that it be interpreted as a space-time diagram, with distance measured vertically and time horizontally. The wall would then be reinterpreted as a piston in a long tube which is being very rapidly withdrawn, in some prescribed manner, and emits expansion waves, each traveling with constant velocity along the lines indicated in the figure as Mach lines. The similarity is analytic as well as geometric; thus the equation of any Mach line is $y - x \tan \vartheta = \text{const}$, and the value of the constant is functionally related to, say, β, and hence

$$\beta = f[y - x \tan (\beta + \theta)] \tag{3-6}$$

Eq. 3-6 and 3-5 are directly analogous to the equations for a one-dimensional progressive wave which Earnshaw derived in 1860, namely,

$$q = f[x - (a + q)t], \qquad a - \tfrac{1}{2}(\gamma - 1)q = a_\infty \tag{3-7}$$

The same term "simple wave" came to be used for both phenomena when this analogy was realized. In both cases the solutions were inferred from physical considerations, but it may easily be verified analytically that they solve the appropriate equations of motion.

Notice that the assumption of uniform specific entropy (which follows from that of a uniform undisturbed stream) is essential for the existence of a simple wave. Without it, even if the velocities were constant along Mach lines, the speed of sound would not be constant; hence the Mach lines would not be plane waves, and therefore would not propagate unchanged unless some sort of reflection along the other system of characteristics was permitted.

Fig. E,3c shows what happens when the simple wave theory is applied in order to determine the effect of bends in the wall concave to the stream. There is no difficulty in applying the boundary conditions as before. A unique mathematical solution of the form of Eq. 3-6 (taken with Eq. 3-5) still emerges. But in the corresponding physical flow picture (Fig. E,3c) the Mach lines converge as we follow their course away from the wall. This is because ϑ (as given by Eq. 3-5) is an increasing function of θ; as the flow turns, the direction of the Mach lines turns in the same sense.

As a result, in the region of compression, any two adjacent Mach lines cross each other at some point. The locus of such intersections is the envelope of the Mach lines. This is found to be a cusped figure, shown as a heavy line in Fig. E,3c. Through each point *outside* it there is one Mach line which defines the value of each physical quantity at the point. Through each point *inside* it there are three Mach lines, giving three values for each physical quantity at the point, which is nonsense.

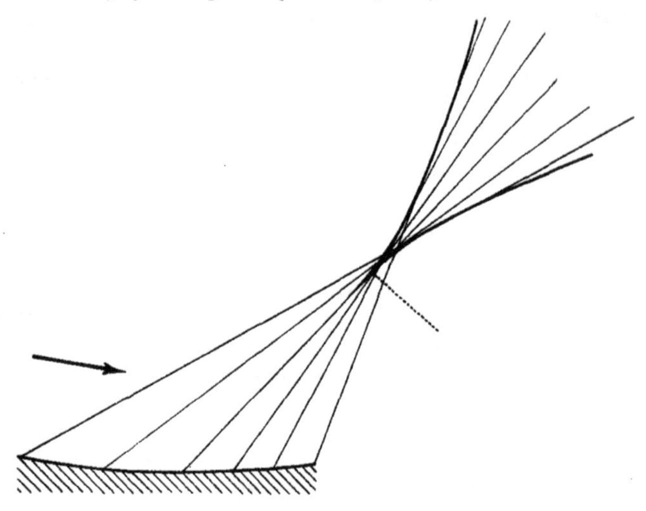

Fig. E,3c. Supersonic flow past a concave wall.

We may pause before discussing what really happens to remark that, whatever it is, provided that the flow remains supersonic, its influence must be confined to the influence domain of the cusp, and in particular the simple wave solution must remain accurate below the dotted line in Fig. E,3c, which represents the characteristic of the other family which passes through the cusp.

But the step to determining what really happens is easily taken in the light of the theory of the formation of shocks in one-dimensional unsteady flow (III,C), especially since the flow plane in the present problem has qualitatively identical properties to the space-time plane in the other problem. The steady two-dimensional flow pattern may be thought of as formed progressively, as the influence of the wall spreads out into the fluid at the speed of sound. The large gradients at the base of the cusped figure are resisted by viscous stresses normal to the cusp. These keep the characteristics from crossing, but they will still get closer and closer, and so need larger and larger viscous stresses to keep them apart. The result is a shock which originates at (or at least very close to) the cusp, with zero strength (i.e. at the Mach angle to the local flow, and thus parallel to the cusp), and whose strength increases as its path is followed away from the wall. The beginning of the path, at least, lies between the two arms of the cusped figure, and near it one may confidently predict that the flow is given approximately by assuming that the physical quantities are still constant along those parts of Mach lines which lie between the wall and their intersection with the shock. (That there is only one such part of a Mach line through a point within the cusped figure can be seen by examination of Fig. E,3c.) The reason for expecting this to be so is that the flow so postulated satisfies the equations of motion under the conditions of zero vorticity and uniform entropy; these conditions are not satisfied exactly behind a shock, but are satisfied to good approximation behind a weak shock; and therefore the flow is possible if there is some curve along which the shock can lie such that the Rankine-Hugoniot shock conditions will correctly link the two values of each physical quantity at the curve (as calculated from the two portions of simple wave, borne on Mach lines entering the cusped figure respectively across one branch or the other). This is not exactly true, unless some propagation from the shock along characteristics of the other family than that shown in Fig. E,3c is permitted; this propagation may be thought of as a result of reflection of the rear simple wave at the shock and transmission of the front one through it; but, again, it may be neglected to a good approximation. On these approximations the position of the shock within the simple wave may be calculated by a method due to Friedrichs [*19*] which is described in another connection in Art. 4. But the problem of the formation of a shock in the midst of an envelope of Mach lines is not further discussed in this section, since the study of the flow past a wall which bends in the direction concave to the flow is not of frequent application to supersonic airfoil theory except when the bending is *sudden*, as at the leading and trailing edges.

But some idea of the behavior of the flow at, say, the leading edge, in the case where the necessary deflection of the stream there is concave to the flow, can be obtained by making the curvature of the wall tend to infinity in the flow studied in the last paragraph (and illustrated in Fig. E,3c), while the length of wall with this curvature is made to tend to zero. In the limit the region covered thrice by Mach lines is bounded (see Fig. E,3d) by two straight lines, of which that on the right is parallel to the Mach lines in the undisturbed stream. Again the shock must occur somewhere in the midst of this region; and only parts of Mach lines between the wall and the shock are of physical significance (other parts of Mach lines are shown dashed in Fig. E,3d). But now the shock is of finite nonzero strength from where it starts at the wall; the portion where it grows from zero strength has vanished in the limit, and now the strength decreases as distance from the wall increases, in a manner to be investigated in Art. 4. However, provided that the shock strength in question is not too large, it may still be a good approximation to neglect the gain in specific entropy at the shock, and the "reflection" of the rear simple wave at the shock, and so take the flow behind the shock as given correctly by the same simple wave theory which is applicable when all the bends in the wall are convex to the stream.

The simple wave theory of airfoil flow. This approximation, when applied at all such corners which occur, affords a complete approximate

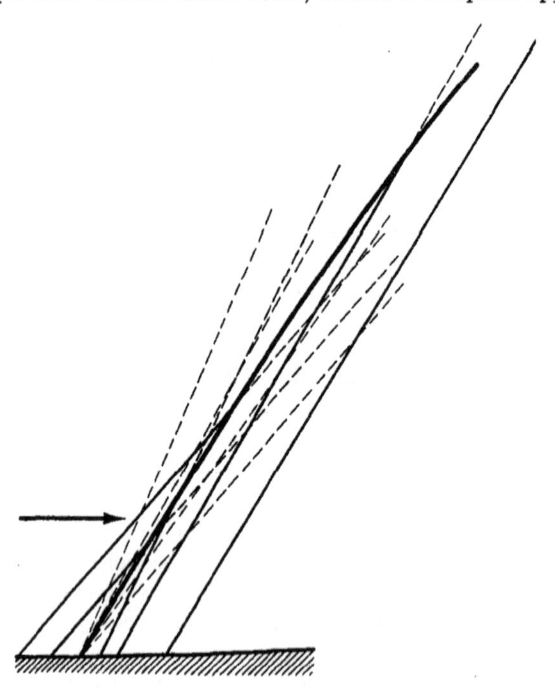

Fig. E,3d. Supersonic flow around a concave corner.

theory of the supersonic flow past an airfoil, which is illustrated in Fig. E,3e for a case where the angle of attack is such that shocks are needed at all four corners. (For larger angles of attack, the shocks in the top left and bottom right-hand corners would be replaced by Prandtl-Meyer expansions.) Between the leading and trailing edge shocks, both above and below the airfoil, there is a simple wave flow. If the direction of motion θ at any point is measured in the positive sense in the upper simple wave, and in the negative sense in the lower simple wave (i.e. in both cases in the sense away from the solid surface), then the local Mach angle β at any point is given in terms of θ by Eq. 3-5. Since θ is known on the airfoil surface, being simply the inclination of the surface to the stream, the values of β on the surface are deduced at once, and hence also the pressure distribution, by the relation

$$p = \frac{p_\infty \left(\dfrac{\sin^2 \beta}{\gamma - \cos 2\beta}\right)^{\gamma/(\gamma-1)}}{\left(\dfrac{\sin^2 \beta_\infty}{\gamma - \cos 2\beta_\infty}\right)^{\gamma/(\gamma-1)}} \tag{3-8}$$

which follows from Bernoulli's equation when the specific entropy is taken as uniform.

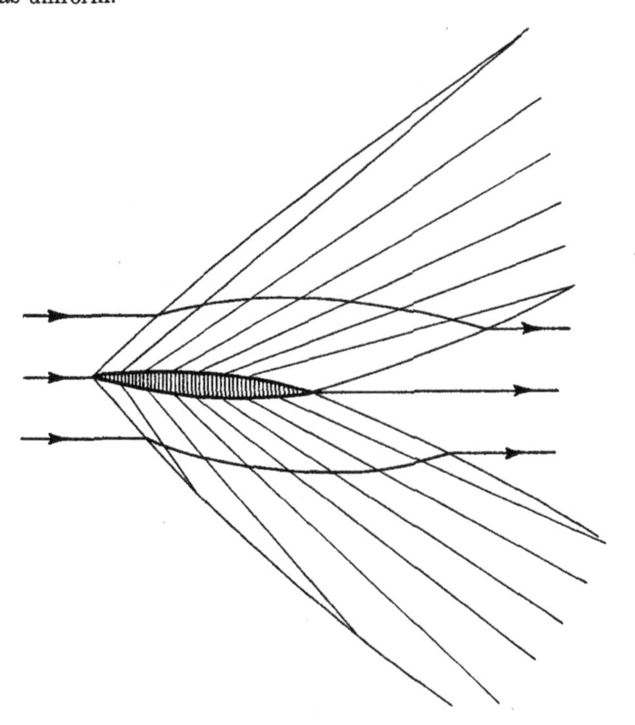

Fig. E,3e. Flow about an airfoil.

The shapes of the shocks remain to be found. No shapes will cause all the Rankine-Hugoniot conditions to be satisfied at the shocks exactly; this is why the simple wave solution is only approximate; but Friedrichs's determination [*19*] of the shock shapes, to within the order of accuracy to which the conditions can be made to apply, is given in Art. 4. This development in the supersonic airfoil theory came late because it is not necessary for the purpose of estimating the surface pressures.

The flow behind the trailing edge shocks must be uniform and parallel to the main stream on the approximate simple wave theory. For, where the streams from above and below the airfoil unite, the pressure must be continuous. If here the direction of motion, measured in the positive sense, is θ, then the pressure is given as a certain function of θ, by Eq. 3-5 and 3-8, for the upper stream, and as the same function of $-\theta$ for the lower stream. These are equal only if $\theta = 0$. The argument applies all along the boundary separating the two streams. On this θ has, and therefore all the physical quantities have, the same values as in the incident stream; hence they have them everywhere behind the trailing edge shock, since quantities are propagated (up to this shock) unchanged along Mach lines. Fig. E,3e, like the previous figures in this article, represents a flow actually calculated from the theory; it includes some representative streamlines. Notice how the stream tube area increases with the rise in velocity along the airfoil surface.

It is fairly easy to estimate the size of the error in the prediction of surface pressure distribution given by simple wave theory. This is because the pressure at the leading edge itself can be calculated exactly from the equations for an oblique shock when the deflection is known. It is true that for a given deflection two alternative shocks are theoretically possible, one stronger than the other (III,E,2.) But it is now fully established experimentally and theoretically that the stronger of the two never occurs except when there is sufficient blockage downstream to make the greater pressure increase necessary in the sense of the steady one-dimensional theory of nozzle flow. (Of course in such a case a region of subsonic flow downstream protests against the resulting low pressure by sending a shock upstream to cause the oblique shock pattern to alter.) In the flow past an isolated airfoil there is of course no blockage, and it is therefore practically certain that the stronger of the two theoretically possible shocks never appears.

In Fig. E,3f and E,3g, the conclusions of the simple wave theory are summarized respectively for $\theta > 0$ and $\theta < 0$. In Fig. E,3f they are critically examined. In this figure the ordinate is the pressure coefficient $C_p = (p - p_\infty)/\frac{1}{2}\rho_\infty q_\infty^2$. The abscissa is M_∞, the Mach number of the incident stream. The curves give the relation between C_p and M_∞ for various positive values of θ, the direction of the tangent to the surface. The plain lines denote the relation given by simple wave theory, from Eq. 3-5 and 3-8. As a comparison, the true value of C_p at the leading edge, as calculated from oblique shock theory, is indicated by a dashed line. The dotted lines will be explained presently. The dash-dotted lines denote the values given by the Ackeret small perturbation theory. (These are included only for the two smallest values of θ in order not to overcrowd the figure, and because the curves for the larger values of θ are simply obtained by scaling up those shown.) The simple wave curves terminate on the left where the magnitude of the deflection first requires the fluid speed to fall below the critical value q^*; no further deflection by a simple wave is possible. Similarly, the curves relating to the oblique shock terminate at the lowest Mach number for which the given deflection is possible.

It is seen that there is remarkably good agreement even for $\theta = 20°$, over the range of Mach numbers for which this deflection is possible by

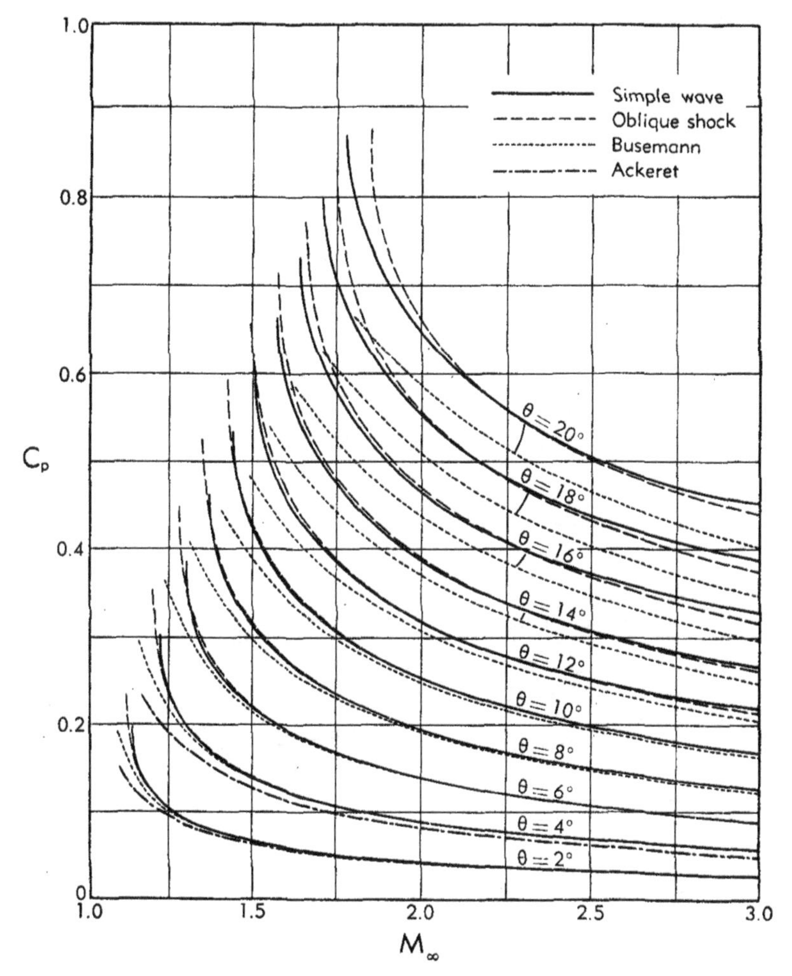

Fig. E,3f. Pressure coefficient produced by flow deflection through an angle θ, plotted against the initial Mach number M_∞.

means of an oblique shock or by a simple wave. For $\theta = 8°$ the curves are practically indistinguishable. This suggests that the expressions for C_p as function of θ, as given by the theories of the simple wave and the oblique shock, might with advantage be expanded in powers of θ, in order to see up to what power they are in agreement. This was first done

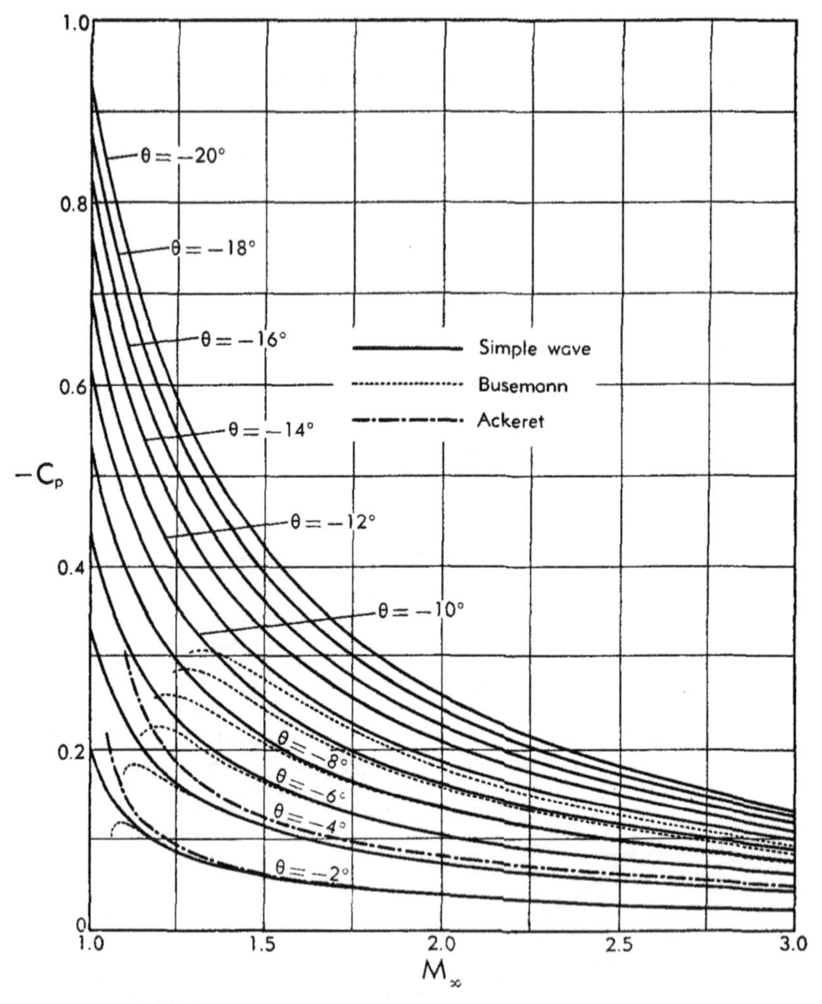

Fig. E,3g. Pressure coefficient produced by flow deflection through a negative angle θ, i.e. convex to the stream.

by Busemann [*20*]. He obtained that, on simple wave theory,

$$C_p = C_1\theta + C_2\theta^2 + C_3\theta^3 + \cdots \tag{3-9}$$

while on oblique shock theory

$$C_p = C_1\theta + C_2\theta^2 + (C_3 - D)\theta^3 + \cdots \tag{3-10}$$

where

$$C_1 = \frac{2}{(M_\infty^2 - 1)^{\frac{1}{2}}}, \qquad C_2 = \frac{(M_\infty^2 - 2)^2 + \gamma M_\infty^4}{2(M_\infty^2 - 1)^2} \tag{3-11}$$

and C_3 and D were given incorrectly, probably as a result of copying errors. Kahane and Lees [*29*] first gave them correctly as

$$C_3 = \frac{M_\infty^4\{[2(\gamma+1)M_\infty^2 + 2\gamma^2 - 7\gamma - 5]^2 - 4\gamma^4 + 28\gamma^3 + 11\gamma^2 - 8\gamma - 3\} + 2(\gamma+1)(3M_\infty^2 - 4)^2}{24(\gamma+1)(M_\infty^2 - 1)^{\frac{7}{2}}} \tag{3-12}$$

$$D = \frac{(\gamma+1)M_\infty^4[(5 - 3\gamma)M_\infty^4 - (12 - 4\gamma)M_\infty^2 + 8]}{48(M_\infty^2 - 1)^{\frac{7}{2}}} \tag{3-13}$$

The asymptotic values of the coefficients as $M_\infty \to 1$ are

$$C_1 \sim \frac{2}{(M_\infty^2 - 1)^{\frac{1}{2}}}, \qquad C_2 \sim \frac{\gamma+1}{2(M_\infty^2 - 1)^2}, \qquad C_3 \sim \frac{(\gamma+1)^2}{3(M_\infty^2 - 1)^{\frac{7}{2}}},$$

$$D \sim \frac{(\gamma+1)^2}{48(M_\infty^2 - 1)^{\frac{7}{2}}} \tag{3-14}$$

These are in agreement with the transonic similarity theory (A,7;D,33) under which Eq. 3-9 is replaced by

$$C_p = \frac{2}{\gamma+1}\{M_\infty^2 - 1 - [(M_\infty^2 - 1)^{\frac{3}{2}} - \tfrac{3}{2}(\gamma+1)\theta]^{\frac{2}{3}}\} \tag{3-15}$$

and Eq. 3-10 by

$$\theta = \tfrac{1}{2}C_p[M_\infty^2 - 1 - \tfrac{1}{4}(\gamma+1)C_p]^{\frac{1}{2}} \tag{3-16}$$

The asymptotic values of the coefficients as $M_\infty \to \infty$ are

$$C_1 \sim \frac{2}{M_\infty}, \qquad C_2 \sim \frac{\gamma+1}{2}, \qquad C_3 \sim \frac{\gamma+1}{6}M_\infty,$$

$$D \sim \frac{(\gamma+1)(5 - 3\gamma)}{48}M_\infty \tag{3-17}$$

These are in agreement with the hypersonic similarity theory (A,6;D,34) under which Eq. 3-8 is replaced by

$$C_p = \frac{2}{\gamma M_\infty^2}\left\{\left[1 + \frac{1}{2}(\gamma - 1)M_\infty\theta\right]^{2\gamma/(\gamma-1)} - 1\right\} \tag{3-18}$$

and Eq. 3-10 by

$$C_p = \frac{1}{M_\infty^2}\left\{\left[4(M_\infty\theta)^2 + \left(\frac{\gamma+1}{2}\right)^2(M_\infty\theta)^4\right]^{\frac{1}{2}} + \frac{\gamma+1}{2}(M_\infty\theta)^2\right\} \tag{3-19}$$

The coefficients C_1, C_2, C_3 are positive for all M_∞ for any gas, but the coefficient D (which is such that $D\theta^3$ is the first approximation to the overestimate of leading edge pressure coefficient given by simple wave theory for $\theta > 0$) changes sign as M_∞ varies. For air ($\gamma = 1.40$), D is negative when $1.25 < M_\infty < 2.54$ and positive outside this range. But even where D is positive it is only a small fraction, say 10 per cent, of C_3.

This goes far toward explaining the remarkable agreement between the solid and broken lines in Fig. E,3f, even for $\theta = 20°$.

The error in the simple wave theory at the leading edge, which has just been investigated, is taken as an estimate of the error at all stations along the surface; there is no reason to suppose that the error would increase markedly from this value, since the shock is at its strongest at the leading edge. This supposition is supported by a more detailed investigation below.

It was not clear a priori that the error should have been of order θ^3 for small θ. It is of course true that the specific entropy change at the shock is known (III,E,2) to be proportional to the cube of its strength for small strengths, and the strength is proportional to θ; in fact the jump s in specific entropy from its value in the main stream (which may be taken as zero) is

$$s \sim c_p \frac{\gamma^2 - 1}{12} \frac{M_\infty^6}{(M_\infty^2 - 1)^{\frac{3}{2}}} \theta^3 \tag{3-20}$$

as $\theta \to 0$. But this jump is not the whole explanation of the difference (approximately $D\theta^3$) between the pressure coefficient obtained by continuous compression and that obtained by a shock, which is partly due to the very fact that the change at the shock is discontinuous, so that the shock is propagated faster than sound.

However, once it is established that both pressure and entropy are in error, on simple wave theory, only by terms of order θ^3, it follows that the same is true of all the thermodynamic variables, including enthalpy. Hence further, by Bernoulli's equation, the same is true of the fluid speed. Thus in all respects simple wave theory is correct to order θ^2, but incorrect to order θ^3.

Busemann's second order approximation. After observing this fact, Busemann [*20*] remarked that it was a reasonable simplification of the problem to neglect all terms of order θ^3, and adopt only the first two terms of Eq. 3-9 and 3-10,

$$C_p = C_1\theta + C_2\theta^2 \tag{3-21}$$

as what has become known as the Busemann approximation to the surface pressure distribution. It is compared with the values given by simple wave theory and oblique shock theory for $\theta > 0$ in Fig. E,3f. It does not remain as close to either as they do to each other, and it diverges from them badly both near the lowest Mach number for which the deflection is possible, and also when M_∞ becomes large. This is because, as Eq. 3-14 and 3-17 show, the radius of convergence of the series of Eq. 3-9 and 3-10 is proportional to $(M_\infty^2 - 1)^{\frac{3}{2}}$ for M_∞ near 1 and to M_∞^{-1} for large M_∞. In Fig. E,3g, the Busemann approximation (3-21) is compared, for $\theta < 0$, with the simple wave theory. The divergence for M_∞ near 1 (due to the small radius of convergence of the series) is of course just as bad as in Fig. E,3f, in spite of the fact that in the present case C_p behaves quite regularly in this region. For the larger expansion angles $-\theta$ the Busemann curves had to be omitted from Fig. E,3g to prevent overcrowding, but the divergence continues to increase. At the higher Mach numbers the absolute errors in the Busemann values for C_p are similar to those for an equal compression angle $+\theta$ (Fig. E,3f).

However, for an ordinary airfoil flying at a Mach number less than 3, where the maximum deflection concave to the stream is such as to permit, with a little to spare, attainment by an oblique shock, the maximum deflection convex to the stream is unlikely to be much greater, and Busemann's approximation can be applied with reasonable accuracy. It is a great improvement on the Ackeret small perturbation theory, in which $C_p = C_1\theta$, which is now seen to be an underestimate of the pressure everywhere except perhaps on portions of surface parallel to the stream. Actually, if we set out to improve on the small perturbation theory after the method of Hantzsche and Wendt (Art. 2), or by a direct expansion (2-19) of the disturbance potential in powers of a thickness parameter, the second approximation obtained is the Busemann approximation, and higher ones are the successive partial sums of the series (3-9) representing the simple wave theory. But, of course, there is no need to do this since the whole exact series can be obtained at once by the simple wave theory, and since in any case both approaches ignore the changes produced by the appearance of a shock.

In using the Busemann approximation (3-21), θ is interchanged at will with $\tan\theta$, the slope of the airfoil surface, since the order of the approximation is not thereby altered. If at "zero angle of attack," when the line joining the leading and trailing edges is parallel to the incident stream, the upper surface is $y = y_u(x)$ and the lower surface is $y = y_l(x)$, then at angle of attack α we therefore put $\theta = -\alpha + y_u'(x)$ on the upper surface and $\theta = \alpha - y_l'(x)$ on the lower surface. Hence Eq. 3-21 becomes

$$C_p = C_1[-\alpha + y_u'(x)] + C_2[\alpha^2 - 2\alpha y_u'(x) + y_u'^2(x)] \tag{3-22}$$

on the upper surface, and

$$C_p = C_1[\alpha - y_l'(x)] + C_2[\alpha^2 - 2\alpha y_l'(x) + y_l'^2(x)] \tag{3-23}$$

on the lower surface. Averaging the difference between Eq. 3-22 and 3-23 along the chord from $x = 0$ (leading edge) to $x = c$ (trailing edge), where $y_u(0) = y_l(0) = y_u(c) = y_l(c)$, to obtain the lift coefficient C_L, we obtain

$$C_L = 2C_1\alpha + \frac{C_2}{c}\int_0^c [y_l'^2(x) - y_u'^2(x)]dx \tag{3-24}$$

Multiplying Eq. 3-22 and 3-23 by the slopes $-\alpha + y_u'(x)$ and $\alpha - y_l'(x)$

respectively, adding, and averaging along the chord to obtain the drag coefficient C_D, we obtain

$$C_D = \frac{C_1}{c}\int_0^c [y_l'^2(x) + y_u'^2(x)]dx + 2C_1\alpha^2$$

$$+ \frac{3C_2\alpha}{c}\int_0^c [y_l'^2(x) - y_u'^2(x)]dx + \frac{C_2}{c}\int_0^c [y_u'^3(x) - y_l'^3(x)]dx \quad (3\text{-}25)$$

The second term in Eq. 3-24 represents the lift at zero angle of attack, which at subsonic speeds would be associated with camber. But we see at once that the influence of camber is different in three important respects from that which is familiar from subsonic flow theory: (1) for an airfoil with a flat lower surface ($y_l(x) \equiv 0$) and curved upper surface, i.e. with *positive* camber, the lift at zero angle of attack is *negative;* (2) if the airfoil which has just been described is modified by *doubling* its camber, so that the upper and lower surfaces coincide, and the airfoil is a curved plate, the lift at zero angle of attack is not doubled but reduced to zero; (3) any effect on lift due to camber is only proportional to its square.

It is clear that any attempt to obtain high maximum lift at supersonic speeds by means of camber would be a mistake. For (1) the camber would have to be negative, which would have a bad influence if the airfoil were ever to be used at subsonic speeds; (2) the deflection on the lower surface, at large angles of attack, would be increased, leading to shock detachment, which is a sort of stall, since it implies rapid flow around the leading edge, and hence probably boundary layer separation on the upper surface; (3) the effect in any case would only be rather small.

The correct method of designing for large maximum lift is to aim at making a large value of α, and hence of the first term in Eq. 3-24, possible. This requires that the distance perpendicular to the stream between the leading and trailing edges be maximized, subject to the deflection of the stream being nowhere too great, and to the wing having sufficient thickness for purposes of strength. This requires a symmetrical airfoil section. It is true that here a large negative value of θ has been avoided equally with a large positive value. But it is in practice true that a strong trailing edge shock can have just as deleterious an effect on lift as a strong leading edge shock, an effect that begins for a deflection considerably below the theoretical maximum (which exceeds that pertaining to the leading edge shock), owing to shock boundary layer interaction.

Thus considerations of maximum lift indicate a symmetrical section as desirable, just as do considerations of low drag according to the Ackeret theory (D,7). For a symmetrical section, the lift coefficient predicted by Eq. 3-24 is $2C_1\alpha$, identical to that predicted by the Ackeret theory. Further, the drag coefficient due to angle of attack is the same, namely, $2C_1\alpha^2$. Only the drag coefficient at zero lift, C_{D_0}, given by the first and last terms of Eq. 3-25, contains a new term due to Busemann's corrected theory of the pressure distribution. This is

$$\frac{2C_2}{c}\int_0^c y_u'^3(x)dx \tag{3-26}$$

This affects the conclusion of the Ackeret theory that the profile of given thickness ratio ϵ whose drag coefficient is least is the diamond shape, symmetrical fore and aft as well as about the stream direction. Taking into account the term (3-26), the profile of least drag is a diamond shape but with the position of maximum thickness a distance $(C_2/2C_1)\epsilon c$ behind the half-chord position. (See Fig. E, 3h, below, for a graph of C_2/C_1.) But this change is hardly of practical significance. In the important case of a section symmetrical about two perpendicular lines, the term (3-26) vanished and the lift and drag coefficients as given by the Ackeret theory are correct even on the more accurate Busemann theory.

As a consequence, the predictions of lift and drag given by the Ackeret theory are well borne out in wind tunnel tests on sharp-nosed sections (which have in practice usually been chosen with this double symmetry) at moderate angles of incidence. The measured drag is greater than the predicted drag by an amount which can reasonably be attributed to skin friction. The lift curve slope is given fairly accurately in cases where the boundary layer is turbulent at the trailing edge. But even in these cases the lift falls away from the theoretical value at higher angles of incidence, before full stalling sets in, as a result of the recompression at the trailing edge on the upper surface beginning farther upstream. This effect also reduces the increase of drag with incidence, and indeed makes it much nearer the theoretical value than would be expected, by cancellation with the increased skin friction drag.

But it must be remembered that it is somewhat fortuitous that the small perturbation theory has such success in these predictions. The actual pressure distribution is given much less accurately by it. But since the pressure coefficient is underestimated all around the body, it is not surprising that for a body with double symmetry the resultant lift and drag are given to greater accuracy. However, the position of the aerodynamic center is sensitive to these errors (for example equal forces on the upper surface near the leading edge and on the lower surface near the trailing edge, though with zero resultant, form a couple) and on the Busemann theory it is not at the half-chord position which the Ackeret theory predicts. Thus the nose-up pitching moment coefficient, taken about the half-chord position, is

$$C_M = \frac{C_1}{c^2}\int_0^c x[y_u'(x) + y_l'(x)]dx - \frac{2C_2\alpha}{c^2}\int_0^c x[y_u'(x) - y_l'(x)]dx + \frac{C_2}{c^2}\int_0^c \left(x - \frac{1}{2}c\right)[y_u'^2(x) - y_l'^2(x)]dx \tag{3-27}$$

and even for sections symmetrical only in the ordinary sense (i.e. about the stream direction) the first and third terms vanish, while the second is proportional to the area S of the section, giving $C_M = (2C_2S/c^2)\alpha$. Since $C_L = 2C_1\alpha$ for symmetrical sections, the position of the aerodynamic center predicted by the Busemann theory is therefore a distance

$$\frac{C_2}{C_1}\frac{S}{c} = \frac{(M_\infty^2 - 2)^2 + \gamma M_\infty^4}{4(M_\infty^2 - 1)^{\frac{3}{2}}}\frac{S}{c} \tag{3-28}$$

ahead of the half-chord position. Here S/c may be interpreted as the mean thickness of the airfoil. The coefficient C_2/C_1, which governs the

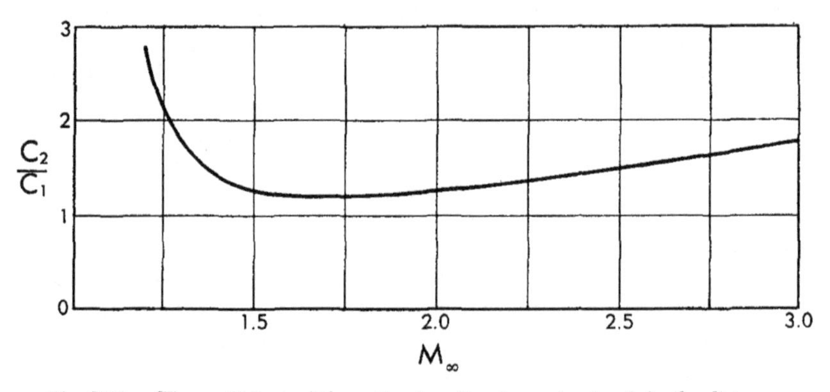

Fig. E,3h. The coefficient of the ratio of sectional area to chord, in the distance of the aerodynamic center ahead of the half-chord position.

variation of aerodynamic center with Mach number for supersonic airfoils, is plotted in Fig. E,3h.

The view taken above is that since the simple wave theory is not accurate (even when the boundary layer is ignored), as a comparison of Eq. 3-9 and 3-10 shows, the further approximation (3-21) might as well be made, since it retains the essential features of the departure from the small perturbation theory, and leads to conveniently simple expressions for the force and moment coefficients. However, for certain airfoil sections (as Epstein [*21*] first showed) one can calculate exact surface pressures (on the assumptions stated at the beginning of the article) which are worth comparing with the Busemann values.

The shock-expansion theory. These sections have a portion near the leading edge consisting of a wedge with both faces plane. When this is so the flow immediately behind the leading edge, say above the airfoil, is uniform, and the values of physical quantities in this flow are given exactly by the oblique shock theory. If the straight portion of the upper surface is AB, where A is the leading edge, then the region of uniform flow is a triangle whose sides are (1) AB, (2) the plane shock AC through A which produces the given deflection, say θ_l, and (3) the Mach line BC through B which is the boundary of the domain of influence of B. Since conditions in ABC are uniform, the Mach line BC is a straight line at the local Mach angle $\beta_l = \csc^{-1} M_l$ to the direction of flow, and so at an angle $\beta_l + \theta_l$ to the incident stream. The pressure coefficient is, say, C_{pl}, given as a function of θ_l and M_∞ by the broken lines in Fig. E,3f.

Now the turning of this uniform flow around the upper surface of the airfoil must be given by the simple wave theory (but, of course with the initial values M_l, θ_l, C_{pl}) provided that the surface lies entirely outside the domain of influence of the point C, where uniformity ceases and (for example) the shock starts to weaken. This may well be the case even if AB is only a small fraction of the chord because the shock AC and Mach line BC make only a small angle with each other, so that C is already a considerable distance downstream.

If the plain lines in Fig. E,3f and E,3g are denoted by $C_{pw}(M_\infty, \theta)$, where w stands for "simple wave," then the pressure on the upper surface behind B will be given, on the hypothesis of the last paragraph, as

$$p = p_l + \tfrac{1}{2}\rho_l q_l^2 C_{pw}(M_l, \theta - \theta_l) \tag{3-29}$$

Hence, subtracting p_∞ and dividing by $\frac{1}{2}\rho_\infty q_\infty^2$, it follows that

$$C_p = C_{pl} + M_l^2(M^{-2} + \tfrac{1}{2}\gamma C_{pl})C_{pw}(M_l, \theta - \theta_l) \tag{3-30}$$

The evaluation of Eq. 3-30 from Fig. E,3f (broken lines) for C_{pl} and Fig. E,3g (plain lines) for C_{pw} requires only a knowledge of M_l, which is deducible from M_∞ and θ_l according to the theory of the oblique shock.

As an example of the results of this method, the flow past an airfoil of "diamond" (i.e. "double wedge," or "rhombus") shape is considered, at variable angle of attack α, for a particular thickness ratio $\epsilon = 0.04$ and Mach number $M_\infty = \sqrt{2} = 1.414$. The results are illustrated in Fig. E,3i. The pressure coefficients on the four sides of the rhombus, with the lift coefficient C_L, drag coefficient C_D and moment coefficient about the half-chord position C_M are shown as plain lines. The approximation of Busemann to each quantity is shown as a dashed line, and that of Ackeret as a dotted line. In the case of C_L and C_D these two coincide (as discussed above), and a dotted line is used. On the other hand C_M, on the Ackeret theory, is identically zero.

It is of some interest to observe what the expansion of Eq. 3-30 in powers of θ and θ_l would be. Omitting all terms of the fourth or higher order in θ and θ_l combined, it is

$$C_p = -D\theta_l^3 + C_1\theta + C_2\theta^2 + C_3\theta^3 \tag{3-31}$$

(No details of this calculation are given, since the reason for Eq. 3-31 becomes clear after the consideration of the flow as a whole to be given in Art. 4.) This supports the statement made above that the error in the simple wave theory was well estimated by its approximate leading edge value $D\theta_l^3$.

When the airfoil surface immediately behind the leading edge is curved, the theory just described (which one may call the shock-expansion theory or Epstein theory) is of course inaccurate, since the shock then starts to weaken from its inception, by interaction with Mach lines, and hence

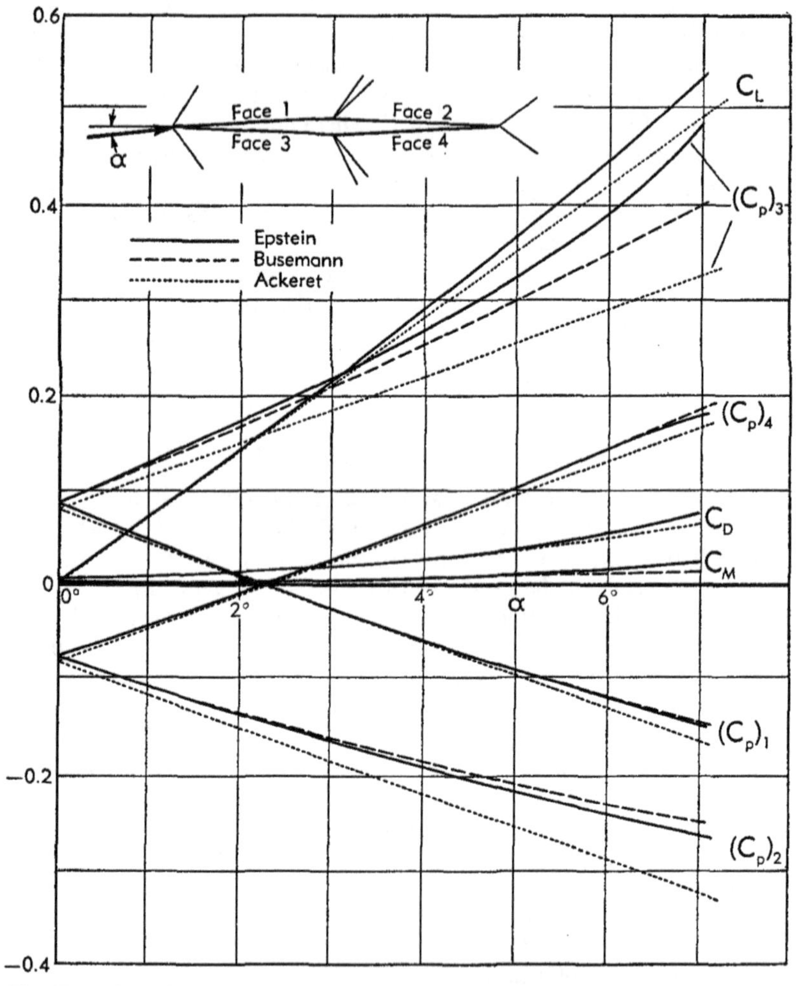

Fig. E,3i. Lift, drag, and moment coefficients, with pressure coefficients on the four faces, for a double wedge airfoil of thickness ratio 0.04 at Mach number 1.414, plotted against angle of attack.

there is no region of uniform specific entropy in which a simple wave theory can be exactly true. However the approximation (3-31) remains true if terms of the fourth order in θ and θ_l are neglected. This is because, though the shock strength is of order θ_l, its gradient (which produces the entropy gradient and hence, by a general theorem on steady flow (III,A), vorticity) is only of order θ_l^2. This in turn is because the shock makes a small angle, of order θ_l, with the Mach lines behind it, so that unit length of shock is only weakened by characteristics emanating from a length of surface of order θ_l, in which θ only decreases by an amount of order θ_l^2.

Thus the errors due to vorticity in Eq. 3-30, where θ_l represents the deflection at the leading edge when this is concave to the stream (and is zero when the deflection is convex so that no shock is formed), are of the order of the fourth power of the maximum deflection. A theory taking into account these errors to a first approximation was already produced by Donov [*22*] in 1939; however, only a review of Donov's paper is available to the present writer. A more systematic theory has recently been given by Thomas [*23*]. Translating his ideas into the notation of the present article, he supposes the pressure coefficient at any point expanded in a series

$$C_p = c_0(\theta) + (s - s_l)c_1(\theta) + (s - s_l)^2c_2(\theta) + \cdots \tag{3-32}$$

to take into account that the pressure is not the same function of the deflection on every streamline (i.e. line on which the specific entropy s is uniform). Then the function to be determined is $c_0(\theta)$, the pressure coefficient on the surface (on which, since it is a streamline, the entropy takes its leading edge value s_l uniformly). Thomas shows that, if as an approximation the series (3-32) is terminated by the term in $(s - s_l)^n$, then the equations of motion and conditions at the shock lead to an ordinary differential equation of the nth order for the required function $c_0(\theta)$. Doubtless only the case $n = 1$ could ever be worked out. Then the equation involves explicitly the curvature at the leading edge, since it is this which determines the extent of the entropy gradient. As Thomas' equations stand there appear to be insufficient initial conditions to determine a solution uniquely, but this deficiency could probably be remedied.

However, such work is irrelevant to the practical conclusions of this article, which are as follows:

1. Under the conditions stated at the beginning of the article, the "simple wave" theory of the surface pressures (which does not need to take any account of the shocks) is, as Fig. E,3f shows, perfectly adequate for all practical purposes, up to a Mach number of 3.

2. For *most* purposes the further simplification due to Busemann is adequate. Fig. E,3f and E,3g show in detail when this is so, and Fig. E,3i illustrates the fact in a particular case. This Busemann approximation implies some interesting conclusions concerning lift, drag, and moment which are displayed in Eq. 3-22 to 3-28 and in the discussion accompanying them.

3. There is some evidence that, even though at very high Mach

numbers the simple wave theory becomes inaccurate, the shock-expansion theory, even for shapes for which it is not exact, may be considerably better.

E,4. Supersonic Two-dimensional Airfoil Theory: Complete Flow Pattern. The subject of this article, the determination of the complete flow pattern when a uniform supersonic stream flows steadily and two-dimensionally past an airfoil, is, on the whole, of general scientific, rather than immediate practical, significance. This is not because in realizations of the problem the surface pressures are the only quantities of importance; for clearly flow conditions, almost anywhere, resulting from the disturbance to the stream made by a wing may be of importance in determining the reaction on other aircraft components or on measuring instruments. Nor is it because the theory neglects the action of viscosity and conduction of heat (principally in the boundary layers and wake), since after the theory has completed its prediction of the flow pattern it is comparatively easy to sketch in the modifications due to these influences. It is rather because (1) in flight the flow is very different from a two-dimensional flow—except on and near the wing surface; (2) in wind tunnel experiments the flow is not at all as though the incident stream were infinitely wide—except on and near the wing surface; indeed most of the complete flow field is cut off by the walls, and the rest modified by reflections.

Nevertheless, the theory is here set out, and for the following reasons. (1) It represents a first step toward the determination of complete flow patterns for more complex problems of supersonic flow (e.g. the analogous problem with three-dimensional axisymmetry) which represent experimental conditions much more closely. (2) Even in this problem no exact solution is possible, but the various approximate approaches may be checked against one another. Since they agree, confidence in all is increased, and in more complex problems one may choose among them the most convenient approach. This increase in confidence is desirable because an infinite flow field is considered and therefore errors can pile up as a wave progresses; also certain theories of phenomena of this kind (involving decaying shocks) assume uniform specific entropy, while others take into account entropy changes and get the same answers, but get different answers when they neglect entropy changes. This state of affairs demands careful study. (3) The theory is closely connected with problems in unsteady flow involving decaying shocks. The connection is greatest with one-dimensional problems, as was noted in Art. 3, but some hope arises of extending the theory to decaying spherical shocks produced by explosions. (4) The contrast with subsonic flow patterns can be fully investigated in this case. It is of especial interest, since a supersonic problem is determined when the upstream state of the fluid is given, to find out to what extent the fluid returns (as it would in subsonic flow) to that state. It is also a question whether the work done by the airfoil against drag all goes (ultimately) into heating up the fluid behind the airfoil, or whether (as will be shown to be the case) some of this energy is transported to an indefinite distance from the line of flight. (5) Finally, as will be seen in Art. 5, the theory is useful in the problem of determining the influence of a blunted leading edge on the flow past an airfoil.

Analysis of the decay of shocks which are not so strong that the dissipation of energy within them is a significant fraction of the internal energy of the fluid was first given, in a special problem of unsteady one-dimensional motion, by Chandrasekhar [*24*], and in one of steady two-dimensional motion by the author [*25*]. A systematic theory for general problems of both these types was later produced by Friedrichs [*19*].

The Friedrichs theory. When applied to the supersonic two-dimensional flow past airfoils this Friedrichs theory is an approximation to the "simple wave theory" which was stated, but not completely worked out, in Art. 3. It remained to determine the front and rear shocks. When this has been done the "simple wave" theory states that the flow between them (either above or below the airfoil) is the simple wave, at the entropy of the undisturbed stream, which is compatible with the boundary condition at the surface. Behind the rear shocks, the fluid returns to the state of the incident stream.

It is possible, as will be shown forthwith, to determine uniquely, on this theory, the positions of the front and rear shocks which will cause *one* of the Rankine-Hugoniot relations for stationary oblique shocks, say that between slope of shock and deflection θ, to be satisfied *exactly* all along it. Then it was seen in Art. 3 that the conditions on the pressure change are not satisfied exactly, the error being of order θ^3 for small θ; and the same is true of the change in any of the physical quantities. Since this is so, Friedrichs simplifies the calculation by neglecting θ^3 throughout. Therefore, the Friedrichs theory is the Busemann theory extended to include the determination of the shocks and the whole flow between them. The Friedrichs theory will first be displayed, and then critically examined and amplified.

In the simple wave springing from, say, the upper surface, the equation of the Mach line on which the deflection takes the value θ is

$$y - x \tan(\beta + \theta) = Y(\theta) \tag{4-1}$$

where β is given in terms of θ by Eq. 3-5, and $Y(\theta)$ is a known function, namely, the value of the left-hand side at the point on the airfoil surface where the deflection of the stream is θ. (It is also the value of y, always negative, where the characteristic produced meets the line $x = 0$ normal to the stream through the leading edge.) With this value for $Y(\theta)$, Eq. 4-1 holds for all points (x, y) in the simple wave. Now the angle η which the front shock makes with the direction of the incident stream at a point where it produces a deflection $\theta > 0$ (and therefore meets the characteristic (4-1)), and also (for $\theta < 0$) the angle η which the rear shock makes with the stream direction at a point where it produces a deflection $-\theta$ (and therefore meets the characteristic (4-1), converting the values of the physical quantities thereon into those for the undisturbed stream) are given by the same function $\eta(\theta)$, as the theory of the shock polar diagram (III,E,2) shows. Hence to determine the shocks (as suggested above) so that this relation between η and θ is satisfied exactly, one must solve

$$\frac{dy}{dx} = \tan\,[\eta(\theta)] \tag{4-2}$$

where y is related to x by Eq. 4-1, to obtain the relation between x and θ on each shock. Eq. 4-1 will then complete the specification of the curves in terms of the parameter θ.

Eq. 4-1 and 4-2 give the linear differential equation

$$\{\tan\,(\beta + \theta) - \tan\,[\eta(\theta)]\}\frac{dx}{d\theta} + x\frac{d}{d\theta}\tan\,(\beta + \theta) = -Y'(\theta) \tag{4-3}$$

for the value of x on the front shock where the deflection by the shock is θ, and also for the value of x on the rear shock where the deflection by the shock is $-\theta$. An explicit solution of Eq. 4-3 could be written down with little trouble. But, for reasons explained above, there is little gained by including terms of order θ^3 in the calculations; hence Friedrichs simplifies the matter by neglecting these. Now, neglecting terms in θ^3, we have, for the slope of the Mach lines $\tan\,(\beta + \theta)$ and the slope of the shock $\tan\,[\eta(\theta)]$, the equations

$$\begin{aligned} \tan\,(\beta + \theta) &= \frac{1}{(M_\infty^2 - 1)^{\frac{1}{2}}} + 2A\theta + B\theta^2 \\ \tan\,[\eta(\theta)] &= \frac{1}{(M_\infty^2 - 1)^{\frac{1}{2}}} + A\theta + C\theta^2 \end{aligned} \tag{4-4}$$

where

$$\begin{aligned} A &= \frac{\gamma + 1}{4}\frac{M_\infty^4}{(M_\infty^2 - 1)^2}, \qquad B = \frac{(\gamma + 1)M_\infty^4}{2(M_\infty^2 - 1)^{\frac{7}{2}}}(\gamma M_\infty^2 + 1) \\ C &= \frac{(\gamma + 1)M_\infty^4}{32(M_\infty^2 - 1)^{\frac{7}{2}}}[(\gamma + 1)M_\infty^4 + 4(\gamma - 1)M_\infty^2 + 8] \end{aligned} \tag{4-5}$$

Some remarks on the value of C are needed before proceeding further. It has been given wrongly in the leading papers on the present theory. Thus Lighthill [*25*] states that a shock bisects the directions of the Mach lines ahead of and behind it if terms of order θ^3 be neglected; this means that $\eta(\theta)$ is the mean of $\beta + \theta$ and β_∞, which is obviously true, by Eq. 4-4, if θ^2 be neglected, but is actually false if only θ^3 be neglected. Also, Friedrichs [*19*], in his expansion for $\eta(\theta)$, which he calls β, (Eq. 14-5 of the cited paper) gives a quite incorrect coefficient of θ^2; he appears to have erred by applying Prandtl's relation, between the normal components of velocity ahead of and behind the shock, in the form which is appropriate only when there is no tangential component. Actually this coefficient C of θ^2 in Eq. 4-4 for the shock slope tan $[\eta(\theta)]$ has one rather surprising property. This is that, when generalized from a perfect gas with constant specific heats to a fluid with arbitrary thermodynamic properties, it does not depend simply on M_∞ and on the shape of the isentrope (in the pressure-volume diagram) through the state of the incident stream, but also contains a term proportional to the rate of change of pressure with entropy (say at constant volume). Thus, though the expressions for all the flow quantities behind the shock, neglecting θ^3, are uninfluenced by the change in specific entropy at the shock, this is not true of the *slope* of the shock. Martin [*26*], in treating problems of one-dimensional unsteady flow (which are equivalent to our hypersonic case) to a second approximation, gives the equation corresponding to Eq. 4-5 incorrectly because to simplify the calculation he freely assumes no change of specific entropy at the shock; in consequence his theory of the reflection of expansion waves at shocks is erroneous; in reality, to his order of approximation and for his special value $\frac{5}{3}$ for γ, there is no such reflection. But Eq. 4-5 for C has now been checked independently from first principles by several workers, and also checked against the transonic and hypersonic similarity theories (as was done for C_1, C_2, C_3, and D in Art. 3), and therefore it may be accepted with full confidence.

Eq. 4-3, with the values of Eq. 4-4 substituted in it, has the exact solution

$$x = \frac{1}{A\theta_f^2}\left(1 + \frac{B - C}{A}\theta_f\right)^{-2C/(B-C)} \int_{\theta_f}^{\theta_l} \theta\left(1 + \frac{B - C}{A}\theta\right)^{(3C-B)/(B-C)} dY(\theta) \quad (4\text{-}6)$$

for the front shock. From here onward θ_f, a function of x, is used to signify the value of θ at the front shock (on the simple wave side of it) at a distance from the leading edge whose component measured downstream is x; then $x = 0$ when θ_f takes its leading edge value θ_l, which has been used in obtaining Eq. 4-6. The same equation holds, with θ_r, the value of θ at the rear shock, written for θ_f, θ_t for θ_l, and with the integral increased by the constant

$$cA\theta_t^2\left(1 + \frac{(B - C)\theta_t}{A}\right)^{2C/(B-C)}$$

where c and θ_t are the values of x and θ at the trailing edge.

It follows at once from Eq. 4-6 that, as $\theta_f \to 0$, x tends to infinity like a multiple of θ_f^{-2} (since the integral tends to a nonzero limit). The same result is true concerning θ_r. Thus on both shocks the shock strength, as indicated by the deflection, varies as the *inverse square root* of the distance from the leading edge when this distance is large compared with the airfoil chord. It follows also, by Eq. 4-1 and Eq. 4-4, that, since, e.g., x is proportional to θ_f^{-2}, $y - x(M_\infty^2 - 1)^{-\frac{1}{2}}$ is proportional to θ_f^{-1}, and hence to $x^{\frac{1}{2}}$, for large x. Thus each shock is asymptotically a parabola with axis at the Mach angle $\beta_\infty = \tan^{-1}[(M_\infty^2 - 1)^{-\frac{1}{2}}]$ to the incident stream. The two shocks become farther and farther apart, as their course is followed outward, like the square root of the distance from the leading edge. The conclusion from this that their slope differs from that of the Mach lines in the undisturbed stream by an amount asymptotically proportional to $x^{-\frac{1}{2}}$, and so to the strength, is a check on the above arguments.

These conclusions had previously been derived by the present author [*25*] from the fact that the shock direction bisects (at least to a first approximation) the angle between the Mach lines in the undisturbed stream and those in the simple wave behind the shock. Since, at great distances, the Mach lines are "as if" concurrent (because, at least, they all intersect the airfoil surface, which is small compared with the distances involved), the shocks must be parabolas with axis parallel to the Mach lines in the undisturbed stream and focus at the point where the Mach lines in the simple wave approximately concur, by the property of a parabola which causes it to be used as a mirror section.

For giving quantitative conclusions Eq. 4-6 and its companion equation for the rear shock may be approximated (without loss of accuracy greater than has been incurred already) by neglecting factors differing from unity by terms of order the square of any deflection occurring. One obtains

$$x = \frac{1}{A\theta_f^2}\int_{\theta_f}^{\theta_l}\left(\theta + \frac{3C - B}{A}\,\theta^2\right)dY(\theta) - \frac{2C}{A^2\theta_f}\int_{\theta_f}^{\theta_l}\theta dY(\theta)$$

$$x = \frac{1}{A\theta_r^2}\int_{\theta_r}^{\theta_t}\left(\theta + \frac{3C - B}{A}\,\theta^2\right)dY(\theta) + \frac{c\theta_t^2}{\theta_r^2} - \frac{2C}{A^2\theta_r}\int_{\theta_r}^{\theta_t}\theta dY(\theta) \tag{4-7}$$

By Eq. 4-1, $Y(\theta)$ may be regarded (to this approximation) as the value of

$$y_u(x) - x\left[\frac{1}{(M_\infty^2 - 1)^{\frac{1}{2}}} + 2Ay_u'(x)\right] \tag{4-8}$$

at the point where $y_u'(x) = \theta$. (In this article, it is not desired, as it was in Art. 3, to vary the angle of attack. Hence the equations of the upper and lower surfaces will be taken as $y = y_u(x)$, $y = y_l(x)$, where y is measured normal to the stream, even when the angle of attack is not zero.) Eq. 4-7 and 4-8, with 4-1, complete Friedrichs's analytical approximation to the flow pattern, by specifying parametrically the position of the shocks and giving the value of the stream deflection at any point of each.

It is of particular interest to determine the coefficients in the asymptotic behavior of the shocks as $x \to \infty$. This is achieved by letting θ_f, $\theta_r \to 0$ in Eq. 4-7. One obtains for the front shock that

$$\theta_f \sim \left[\frac{y_p}{(M_\infty^2 - 1)^{\frac{1}{2}}} + K \int_0^{x_p} y_u'^2(x)dx\right]^{\frac{1}{2}} (Ax)^{-\frac{1}{2}} \tag{4-9}$$

where subscript p signifies conditions at what Friedrichs calls the "peak" of the upper surface of the airfoil, namely, the point where the flow is parallel to the undisturbed stream, and where

$$K = \frac{3C - B}{A(M_\infty^2 - 1)^{\frac{1}{2}}} + A - 1 = \frac{(5\gamma - 3)M_\infty^4 - 4(\gamma - 1)M_\infty^2}{8(M_\infty^2 - 1)^2} > 0$$

Similarly, for the rear shock,

$$-\theta_r \sim \left[\frac{y_p - y_t}{(M_\infty^2 - 1)^{\frac{1}{2}}} - K \int_{x_p}^{c} y_u'^2(x)dx\right]^{\frac{1}{2}} (Ax)^{-\frac{1}{2}} \tag{4-10}$$

The strengths $\Delta p/p$ of the shocks for $x \to \infty$ are obtained by multiplying Eq. 4-9 and 4-10 respectively by $\gamma M_\infty^2(M_\infty^2 - 1)^{-\frac{1}{2}}$. It is of some interest to note that the dependence of the strength of the front shock on y, the distance from the airfoil perpendicular to the stream, is to the first order as $2\gamma \sqrt{[y_p/(\gamma + 1)y]}$, which is independent of Mach number. The asymptotic shock shapes, by Eq. 4-1 and 4-4, are

$$y - \frac{x}{(M_\infty^2 - 1)^{\frac{1}{2}}} \sim 2(Ax)^{\frac{1}{2}} \left[\frac{y_p}{(M_\infty^2 - 1)^{\frac{1}{2}}} + K \int_0^{x_p} y_u'^2(x)dx\right]^{\frac{1}{2}} \tag{4-11}$$

for the front shock and

$$y - \frac{x}{(M_\infty^2 - 1)^{\frac{1}{2}}} \sim -2(Ax)^{\frac{1}{2}} \left[\frac{y_p - y_t}{(M_\infty^2 - 1)^{\frac{1}{2}}} - K \int_{x_p}^{c} y_u'^2(x)dx\right]^{\frac{1}{2}} \tag{4-12}$$

for the rear shock. Thus, to a first approximation, the shocks form part of the same parabola if the angle of attack is zero, since then $y_t = 0$. To a second approximation this is not true, according to Eq. 4-11 and 4-12, since the additional terms within brackets have opposite signs.

Friedrichs [*19*] was careful not to suggest that the coefficients in Eq. 4-9 to 4-12 are correct, neglecting cubes of disturbances, but described them merely as approximations. Actually (as shown in the critical discussion below) the coefficients in Eq. 4-9 and 4-11 are correct in this sense, but the second order terms in the coefficients in Eq. 4-10 and 4-12 are wrong for really large x, and will be emended. In other words the Friedrichs theory gives the asymptotic behavior of the *rear* shock correct only to a *first* approximation, although it gives that of the front shock correct to a second approximation.

The variation of pressure throughout the field is given in terms of the variation of the deflection θ by Eq. 3-21. As $x \to \infty$, the variation of C_p between the shocks, along a line x = const perpendicular to the stream, is therefore given (using Eq. 4-1 and 4-4) by

$$\frac{\partial C_p}{\partial y} \sim C_1 \frac{\partial \theta}{\partial y} \sim C_1 \frac{1}{2Ax} = \frac{4(M_\infty^2 - 1)^{\frac{3}{2}}}{(\gamma + 1)M_\infty^4} \frac{1}{x} \tag{4-13}$$

Thus, for a given Mach number, the pressure gradient between the shocks becomes independent of the airfoil shape, and inversely proportional to the distance from the airfoil, when this distance is large enough. The pressures at the two shocks, and also their distance apart, are sensitive to the airfoil shape, but the uniform pressure gradient between the shocks is unaffected by it at large distances from the surface. In

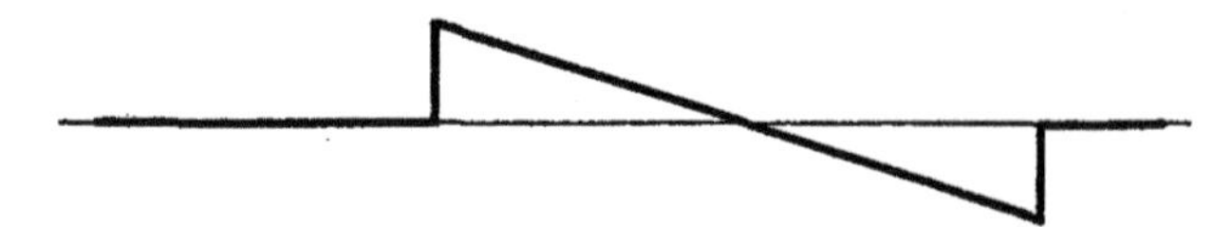

Fig. E,4a. Pressure distribution in an N wave.

Fig. E,4a the pressure between the shocks is plotted for a large constant value of x with the front shock on the left. Owing to the shape of this curve the configuration is called an N wave. For given x and M_∞ the heights of the uprights of the N depend on airfoil shape but the slope of the diagonal portion is uniquely determined. Similar conclusions apply to a plot of pressure for a large constant value of y.

Effects of entropy variation. It might be supposed that to obtain a higher approximation than the Friedrichs theory, which would give some insight into the flow behind the rear shock, would be very laborious, since the variation in entropy in the simple wave would have to be taken into account. Actually it is not so, owing to the following useful fact. Variations of entropy may, to a first approximation, be neglected when variations of the pressure p or the direction of motion θ are being evaluated. Only in deducing other quantities, e.g. the density ρ or the fluid speed q, need the entropy variations be taken into account.

Put more precisely, the principle states that a difference in p or θ between adjacent points is only influenced by the entropy difference to an amount of the order of this entropy difference *multiplied by* the difference in p or θ. The most rigorous proof is from the equations of the characteristics of steady plane flow with nonuniform entropy (G,18); these state that

$$dp \pm (\rho q^2 \tan \beta) d\theta = 0 \qquad \text{when} \qquad \frac{dy}{dx} = \tan (\theta \pm \beta)$$
$$ds = 0 \qquad \text{when} \qquad \frac{dy}{dx} = \tan \theta \tag{4-14}$$

a form probably first given by R. E. Meyer [27]. If Eq. 4-14 are solved with ρ, q, and β given as functions of p by Bernoulli's equation with a *uniform* value for the specific entropy s, then the error is of the order of the error in s multiplied by the change in θ (not simply of the order of the error in s itself as it would be if any quantity other than p or θ were considered). In particular, in the flow behind a shock the error is of the order of the fourth power of the shock strength.

Physically the result follows, as far as the pressure is concerned, from the fact that at any point the local equations are those of sound, and that pressure is propagated acoustically independently of small vorticity or variations in entropy. The result for the direction θ depends more explicitly on the fact that the flow is steady.

Now the result does not mean that the simple wave theory is true to a higher order than might first have been supposed; for the result applies only to the equations of motion, not to the conditions which must be satisfied at the shocks. These are incompatible with the conditions in the simple wave. It was found in Art. 3, to a first approximation, just what the discrepancy was, in terms of p and θ. At any point immediately behind the front shock, by Eq. 3-9 and 3-10, the pressure is less by approximately $\frac{1}{2}\rho_\infty q_\infty^2 D\theta^3$ than what it would be on the simple wave theory. This causes an additional wave, superposed on the simple wave, and propagated along the second system of Mach lines $dy/dx = \tan(\theta - \beta)$ (see Fig. E,4b); this additional wave may be regarded as the reflection of the simple wave at the front shock. The interaction between the simple wave and its reflection may here be neglected as being of the order of the product of their strengths, and so of the order of the fourth power of disturbances.

Using subscript s for values of quantities in a simple wave, the equation $dp_s - (\rho q^2 \tan\beta)d\theta_s = 0$ holds everywhere, not merely (4-14) on the Mach lines $dy/dx = \tan(\theta - \beta)$, since on the other Mach lines $dp_s = d\theta_s = 0$. Similarly, using subscript R for values of quantities in the additional pressure and direction fields due to the reflection of the simple wave at the shock, the equation $dp_R + (\rho q^2 \tan\beta)d\theta_R = 0$ holds everywhere. Since we are not interested in quantities of smaller order than θ_R (i.e. than the cube of the disturbances) this may be integrated approximately as

$$p_R + (\rho q^2 \tan\beta)\theta_R = 0 \tag{4-15}$$

Now at the front shock, $p_s + p_R$ is related to $\theta_s + \theta_R$ as in simple wave theory but with the additional term $-\frac{1}{2}\rho_\infty q_\infty^2 D\theta_f^3$, with subscript f (as above) signifying the value at the shock. But to our approximation the value of p corresponding to $\theta_s + \theta_R$ in simple wave theory would be $p_s + (\rho q^2 \tan\beta)\theta_R$. Hence

$$p_R - (\rho q^2 \tan\beta)\theta_R = -\tfrac{1}{2}\rho_\infty q_\infty^2 D\theta_f^3 \tag{4-16}$$

Eq. 4-15 and 4-16 show that, in the reflection of the simple wave at the front shock, the additional pressure p_R and deflection θ_R (constant along the second system of Mach lines) are given in terms of the value $\theta = \theta_f$

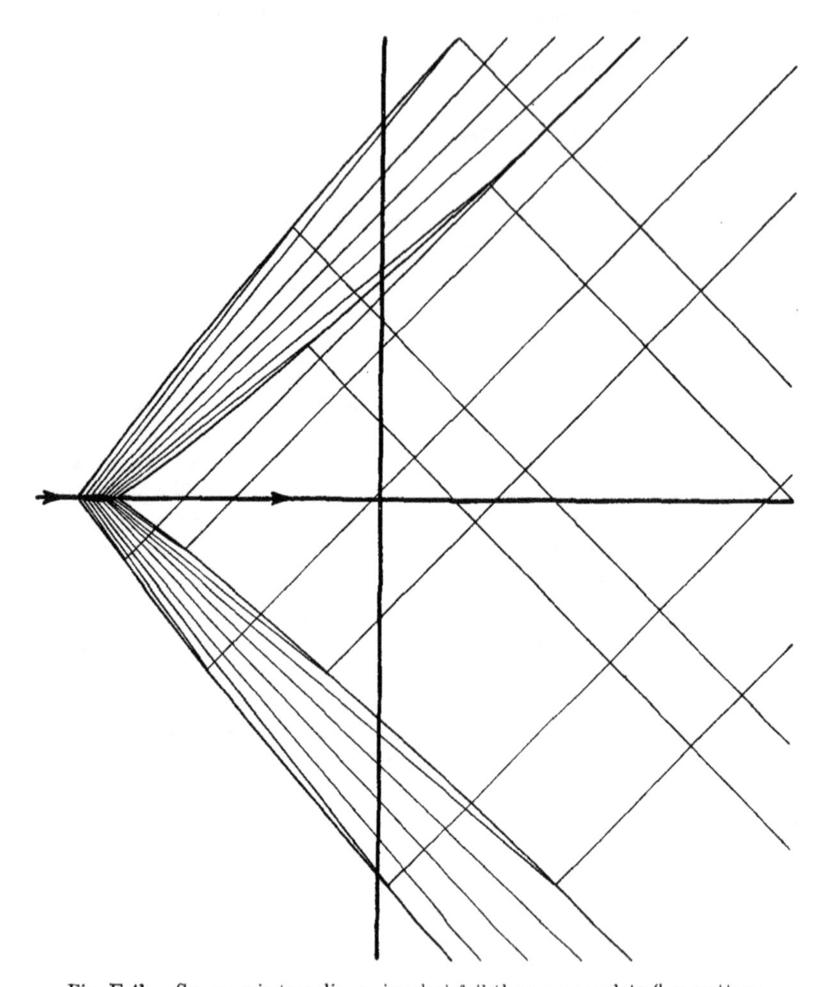

Fig. E,4b. Supersonic two-dimensional airfoil theory: complete flow pattern.

where the Mach line emerges from the front shock as

$$p_R = -\tfrac{1}{4}\rho_\infty q_\infty^2 D\theta_f^3, \qquad \theta_R = \tfrac{1}{4}(M_\infty^2 - 1)^{\frac{1}{2}} D\theta_f^3 \tag{4-17}$$

Part of this reflected wave is in turn reflected at the airfoil surface, after which it goes to modify the simple wave by adding additional third order terms.

In this doubly reflected wave, the pressure is $p_{RR} = p_R$ and the deflection $\theta_{RR} = -\theta_R$. Actually the Mach lines of the second system which hit the airfoil spring from a part of the shock where the deflection does not vary significantly from the leading edge value θ_l. Hence the modification to the simple wave consists approximately of the constant change in pressure $-\frac{1}{4}\rho_\infty q_\infty^2 D\theta_l^3$, which is negative except when $1.25 < M_\infty < 2.54$. At the airfoil surface the additional pressure to that given on simple wave theory is $p_R + p_{RR} = 2p_R = -\frac{1}{2}\rho_\infty q_\infty^2 D\theta_l^3$, agreeing with the conclusions of Art. 3.

Lighthill [28] studied conditions just behind a stationary shock and obtained an exact reflection coefficient for the reflection of the simple waves at the shock. It increases with θ_f initially like θ_f^3 but later somewhat more slowly; for $M_\infty = 2$, the reflection coefficient for the maximum shock deflection is two thirds of that given by the approximate theory above. Of course to higher terms than those retained above there would be an interaction between the simple wave and its reflection at the shock, so that although the theory mentioned gives the exact form of the latter at the shock it cannot predict how it will propagate. Hence the approximate theory, which appears here for the first time, is both adequate and the best that can be done with the amount of labor which the problem merits.

An error in the cited paper [28] should be noted here. Owing to an algebraic slip the limiting value of the reflection coefficient for infinite Mach number was obtained incorrectly as -1 instead of its true value of -0.14, which follows from the work of Eggers and Syvertson [67]. The actual value shows that shock waves are always good absorbers of incident sound waves, even in extreme cases of large Mach number and large deflection angle, and helps to explain the point noted at the end of Art. 3, that the shock-expansion theory is rather good even for hypersonic flow.

We now continue our further analysis of the flow field into the region behind the rear shock, and study the wave which this shock transmits back along the Mach lines of the second system. (It cannot transmit waves behind it along Mach lines of the first system, since these point upstream from it; see Fig. E,4b.) If subscript T signifies conditions in the transmitted wave, the pressure and deflection behind the shock are $p_\infty + p_T$, θ_T. Here (as in Eq. 4-15) $p_T + (\rho q^2 \tan\beta)\theta_T = 0$. Now the pressure and direction ahead of the shock are $p_s + p_R$, $\theta_s + \theta_R$ (where as in all the arguments above the simple wave value p_s includes the reinforcement to the simple wave provided by the doubly reflected wave p_{RR}). Further, the pressure ratio $(p_s + p_R)/(p_\infty + p_T)$ is related to the negative deflection $\theta_s + \theta_R - \theta_T$ exactly as in the ordinary shock equations. In particular this relation differs from the relation given by simple wave theory by a term $-\frac{1}{2}\rho_\infty q_\infty^2 D\theta_r^3$. Approximating to this fact as before, it is easily deduced that

$$p_s + p_R - p_T = p_s + (\rho q^2 \tan\beta)(\theta_R - \theta_T) - \tfrac{1}{2}\rho_\infty q_\infty^2 D\theta_r^3 \quad (4\text{-}18)$$

Hence $p_T = p_R + \frac{1}{4}\rho_\infty q_\infty^2 D\theta_r^3$, i.e.

$$p_T = -\tfrac{1}{4}\rho_\infty q_\infty^2 D(\theta_f^3 - \theta_r^3), \qquad \theta_T = \tfrac{1}{4}(M_\infty^2 - 1)^{\frac{1}{4}} D(\theta_f^3 - \theta_r^3) \tag{4-19}$$

The wave transmitted by the rear shock is thus not identical with that reflected from the front shock; the former is *reinforced* (since $\theta_f > 0$ and $\theta_r < 0$) by an additional wave which can be regarded as the transmission of the simple wave at the rear shock. The mechanism of the process is that a shock, in catching up with acoustic waves, can touch off a small part of their energy as a wave traveling in the opposite direction (the rest of the energy being absorbed in weakening the shock).

The flow behind the rear shocks therefore consists (Fig. E,4b) of a wave transmitted through the upper rear shock, which is propagated along Mach lines pointing downward, and a similar wave transmitted through the lower rear shock, which is propagated along Mach lines pointing upward. Both waves constitute third order effects and are therefore propagated closely according to the small perturbation theory, as far as p and θ are concerned. (Since the entropy variations are of the same order as the pressure waves, the density and fluid speed are *not* obtained from the ordinary small perturbation theory, but from p and the known behavior of s, given by its increment, (Eq. 3-20), at each shock and subsequent uniformity along streamlines.) The values of p and θ propagated along a given Mach line are given by Eq. 4-19, where θ_f and θ_r are the values of θ in the simple wave where the Mach line meets the shocks; these values are given by Eq. 4-7.

The most interesting thing about this tail pressure wave is its breadth. For by Eq. 4-7 and 4-8, the front shock (for example) decays to one half of its initial strength, so that $\theta_f = \frac{1}{2}\theta_l$, in a distance of order θ_l^{-1} chords. This may be taken as a measure of the breath of the tail pressure wave, since the amplitude of waves originating from the shock falls to one eighth of its maximum in this distance. In the limit of infinite distance downstream, both waves have disappeared from the flight path (diagonally) and the pressure and flow direction have returned to their values in the incident stream.

Some attention has been given in the literature to the state immediately behind the trailing edge. At this point the waves are those originating in parts of the shocks still at their initial strengths. Hence the pressure p and the direction of flow θ, measured in the positive sense from the direction of the undisturbed stream (as it has been above for the upper surface but not for the lower surface), are given by

$$p = -\tfrac{1}{4}\rho_\infty q_\infty^2 D[y_u'^3(0) - y_u'^3(c) - y_l'^3(0) + y_l'^3(c)] \tag{4-20}$$

$$\theta = \tfrac{1}{4}(M_\infty^2 - 1)^{\frac{1}{4}} D[y_u'^3(0) - y_u'^3(c) + y_l'^3(0) - y_l'^3(c)]$$

For any airfoil symmetrical about two perpendicular planes, the value of θ given by Eq. 4-20 is zero, since even for nonzero angle of attack $y_u'(0) = y_l'(c)$ and $y_l'(0) = y_u'(c)$. Thus even to a third approximation the streams above and below such an airfoil reunite in a stream locally parallel to the incident stream. This is in marked contrast to subsonic flow past a lifting airfoil, in which the stream is depressed immediately behind the trailing edge. Lighthill [*25*], as a result of a mistake, obtained a nonzero downwash at the trailing edge to this third approximation, but Kahane and Lees [*29*] calculated it by Epstein's method (Art. 3) for the flat plate and double wedge sections, and obtained a fourth order value for the downwash, which they tabulated. It is essentially due to the difference in $(M_\infty^2 - 1)^{\frac{1}{2}}D$ (due to difference in Mach number) at the front and rear shocks. It is always a minute fraction of a degree.

Of course all quantities other than p and θ may be discontinuous at the trailing edge on the present theory, since the entropy increments at the shocks on the two surfaces may not be the same, and this influences all these other flow quantities. A weak vortex sheet (contact discontinuity) may therefore stretch downstream from the trailing edge.

Lift related to downwash. Now that the picture of the flow is complete it is possible to investigate how the airfoil is putting momentum and energy into the fluid by resisting the aerodynamic forces. Besides its independent interest, this provides a check on the work, in both its algebraic and conceptual aspects.

In the case of the lift force, we have to express that this is equal to the rate at which downward momentum ($-\rho v$ per unit mass perpendicular to the incident stream) is transported downstream. Thus the lift $\mathcal{L}$ per unit span is given by

$$\mathcal{L} = -\int_{-\infty}^{\infty} \rho u v dy \tag{4-21}$$

along any line $x = \text{const} > c$ perpendicular to the stream and downstream of the airfoil. The question is: For given x, what ranges of y contribute most to the integral (4-21)?

The answer is at once clear. For any x, the "simple wave" regions between the shocks, both above and below the airfoil, predominate in the momentum integral (4-21). For in those regions the disturbance is of the first order, and elsewhere it is of the third order. Also, when x is many multiples of c, so that the momentum crossing a surface far downstream is considered, the disturbance has been reduced like $x^{-\frac{1}{2}}$ but nevertheless the width of the simple wave region has increased like $x^{\frac{1}{2}}$.

The verification that Eq. 4-21 is correct to the first order for any x, even if the integral be taken only in the simple wave regions, is an easy matter. For consider the value of the integral in the upper simple wave region. To the first order it is

$$-\int_{\theta_r}^{\theta_f} \rho_\infty q_\infty^2 \theta [2Ax + Y'(\theta)] d\theta \tag{4-22}$$

where dy has been substituted in terms of $d\theta$ by Eq. 4-1 and 4-4. By Eq. 4-7 this is, still to a first approximation,

$$-\rho_\infty q_\infty^2 \left[(\tfrac{1}{2}\theta_f^2 - \tfrac{1}{2}\theta_r^2)2Ax + \int_{\theta_l}^{\theta_l} \theta Y'(\theta)d\theta - A\theta_f^2 x + A\theta_r^2 x \right] \quad (4\text{-}23)$$

of which all but the middle term cancels out. Substituting from Eq. 4-8 for $Y(\theta)$, Eq. 4-23 becomes

$$-\rho_\infty q_\infty^2 \int_0^c y_u'(x) \frac{dx}{(M_\infty^2 - 1)^{\frac{1}{2}}} = -\rho_\infty q_\infty^2 (M_\infty^2 - 1)^{-\frac{1}{2}} y_t \quad (4\text{-}24)$$

which is half the lift given by Ackeret's small perturbation theory. The other half comes from the simple wave region below the airfoil.

Thus to the first order it is true that a supersonic airfoil achieves lift by transmitting downwash downstream in two waves stretching out diagonally behind it. On the other hand, of course, a subsonic airfoil transmits downwash downstream in all directions (proportional to the cosine of the direction), and also transmits upwash upstream.

The corresponding investigation carried to the second order of approximation is more complicated but in some ways even more interesting. It will be sufficient here, in order to bring out the significant points, to display the work only for a value of x so large that in the simple wave the asymptotic formulas (4-9) etc. apply, and that also the squares of disturbances may be neglected there. (It is not of course inconsistent to retain the squares of the main disturbances and neglect those of the much smaller disturbances which obtain for very large x.)

Then the integral (4-21), taken over only those values of y within the upper simple wave region, is, to the second order of approximation, still given by Eq. 4-22. But the relations between θ_f, θ_r, and x are now Eq. 4-9 and 4-10. Also the integral of $\theta Y'(\theta)$ between the limits becomes negligible as they approach one another for large x. Thus Eq. 4-22 becomes

$$-\rho_\infty q_\infty^2 Ax(\theta_f^2 - \theta_r^2) = -\rho_\infty q_\infty^2 \left[\frac{y_t}{(M_\infty^2 - 1)^{\frac{1}{2}}} + K \int_0^c y_u'^2(x)dx \right] \quad (4\text{-}25)$$

Adding Eq. 4-25 to the similar term arising from the simple wave below the airfoil, we obtain a second order expression for the lift which ignores any possible contribution from downwash behind the rear shock and may therefore be called the Friedrichs-theory contribution to the lift, $\mathcal{L}_{\text{Friedrichs}}$. This is

$$\mathcal{L}_{\text{Friedrichs}} = \rho_\infty q_\infty^2 \left\{ -\frac{2y_t}{(M_\infty^2 - 1)^{\frac{1}{2}}} + K \int_0^c [y_l'^2(x) - y_u'^2(x)]dx \right\} \quad (4\text{-}26)$$

But the actual lift is as in Eq. 4-26 but with $\frac{1}{2}C_2$ replacing K (Eq. 3-24). Now K is not equal to $\frac{1}{2}C_2$. Hence, either the Friedrichs theory of the flow between the shocks is wrong to the second order, or a significant part of the integral (4-21) comes from behind the rear shock. Actually the latter is true. (Here the parallel with the Stokes-Oseen theory, noted in Art. 1, is very strong.)

This is because, though the disturbances behind the rear shock are of the order of the cube of the maximum disturbances, they form (as has been seen) a wave whose breadth increases as the maximum disturbance decreases, like its inverse first power. Thus the contribution to the integral (4-21), for large x, from the wave transmitted through the upper rear shock, is, by Eq. 4-19,

$$-\rho_\infty q_\infty^2 \tfrac{1}{4}(M_\infty^2 - 1)^{\frac{1}{2}} D \int (\theta_f^3 - \theta_r^3) dy \tag{4-27}$$

But the wave remains unchanged in shape as it is propagated downstream, and a vertical height dy of wave springs from a stretch of shock with horizontal component $dx = \frac{1}{2}(M_\infty^2 - 1)^{\frac{1}{2}} dy$. Hence $(M_\infty^2 - 1)^{\frac{1}{2}} dy$ may be replaced by $2dx$, if the integral in Eq. 4-27 is now taken along the two shocks, and dx in turn may then be expressed in terms of θ_f or θ_r by Eq. 4-7. This gives for Eq. 4-27 (rejecting third order terms), after integration by parts,

$$-\rho_\infty q_\infty^2 \frac{1}{2} D \left\{ \int_0^{\theta_l} 3\theta_f^2 \left[\frac{1}{A\theta_f^2} \int_{\theta_f}^{\theta_l} \theta dY(\theta) \right] d\theta_f \right.$$
$$\left. - \int_0^{\theta_l} 3\theta_r^2 \left[\frac{1}{A\theta_r^2} \int_r^{\theta_l} \theta dY(\theta) \right] d\theta_r \right\} = -\rho_\infty q_\infty^2 \frac{3D}{2A} \int_{\theta_t}^{\theta_l} \theta^2 dY(\theta)$$
$$= -\rho_\infty q_\infty^2 \frac{3D}{2A(M_\infty^2 - 1)^{\frac{1}{2}}} \int_0^c y_u'^2(x) dx \tag{4-28}$$

With this additional term, and with the corresponding one from the wave transmitted through the lower rear shock, the calculated value for the lift integral (4-21) is no longer the erroneous one of Eq. 4-26. The coefficient K is replaced therein by

$$K + \frac{3D}{2A(M_\infty^2 - 1)^{\frac{1}{2}}} \tag{4-29}$$

The final question (whether these considerations of rate of transmission of downward momentum downstream indicate consistency for our complete picture of the flow, and in particular for the Friedrichs theory of the position of the shocks) therefore resolves into calculating whether the function (4-29), of M_∞ and γ, is equal to $\frac{1}{2}C_2$. This is true, as is quickly verified from Eq. 3-10 and 3-12.

This indicates that the Friedrichs theory, which neglects third order terms, is correct to this order although the lift and the rate of transport of downwash downstream calculated from it are inconsistent. For the latter fact is only true because the broad third order pressure wave behind the rear shock makes a second order contribution to the downwash. Precisely the same conclusion is reached when the above calculation is gone through in the more complicated case of general x, rather than in the limiting case $x \to \infty$.

Drag related to downstream flow. When we turn to the problem of correlating drag with downstream flow pattern, we see that it may be put (unlike the problem of lift) in a multiplicity of forms. First we may work with a frame of reference fixed in the airfoil, and equate the drag $\mathfrak{D}$ per unit span to the difference between the flux of longitudinal momentum across planes normal to the stream respectively upstream and downstream of the airfoil, plus the difference between the normal force due to pressure across these planes. This gives

$$\mathfrak{D} = \int_{-\infty}^{\infty} (p_\infty + \rho_\infty q_\infty^2 - p - \rho u^2) dy \tag{4-30}$$

an integral to be evaluated, like Eq. 4-21, for constant $x > c$. On the other hand we may take a frame so that the undisturbed fluid is at rest, and equate the drag to the rate of gain of momentum of the fluid between two such planes (due to the flow pattern moving forward), again plus the difference in normal force. This gives

$$\mathfrak{D} = \int_{-\infty}^{\infty} [p_\infty - p - \rho u(u - q_\infty)] dy \tag{4-31}$$

Also, in this frame, we may equate the rate at which the airfoil does work on the fluid (per unit span), namely $\mathfrak{D}q_\infty$, to the rate of gain of total energy (kinetic and internal) of the fluid between two such planes, minus the sum of the rates of work done by the pressures on the planes, i.e.

$$\mathfrak{D} = \int_{-\infty}^{\infty} \frac{1}{q_\infty} \left\{ p(u - q_\infty) + \rho u \left[e - e_\infty + \frac{1}{2}(u - q_\infty)^2 + \frac{1}{2} v^2 \right] \right\} dy \tag{4-32}$$

Finally, we may work with "wave energy" (in the sense used in acoustics) rather than total energy, i.e. ignore that part of the internal energy associated with compression by the pressure p_∞ of the undisturbed stream. This gives

$$\mathfrak{D} = \int_{-\infty}^{\infty} \frac{1}{q_\infty} \left\{ (p - p_\infty)(u - q_\infty) + \rho u \left[e - e_\infty + \frac{p_\infty}{\rho} - \frac{p_\infty}{\rho_\infty} + \frac{1}{2}(u - q_\infty)^2 + \frac{1}{2} v^2 \right] \right\} dy \tag{4-33}$$

Of the four forms, Eq. 4-31 and 4-33 are to be preferred as concentrating as far as is possible only on departures of quantities from their values in the incident stream. It is therefore especially interesting that, while of course all the four integrals are equal, in the case of Eq. 4-31 and 4-33 the *integrands* are equal everywhere. This follows from Bernoulli's equation

$$e + \frac{p}{\rho} + \frac{1}{2}u^2 + \frac{1}{2}v^2 = e_\infty + \frac{p_\infty}{\rho_\infty} + \frac{1}{2}q_\infty^2 \tag{4-34}$$

Thus the problems of relating the drag to the distributions of flux of additional longitudinal momentum, and of flux of wave energy, are identical.

On the other hand the integrand in Eq. 4-30 differs from that in Eq. 4-31 and 4-33 by q_∞ times $(\rho u - \rho_\infty q_\infty)$. This latter term represents the mass flow across unit area (compared with its upstream value), and its integral over a plane normal to the stream is, of course, zero. Similarly, the integrand in Eq. 4-32 differs from that in Eq. 4-31 and 4-33 by $(p_\infty/\rho_\infty q_\infty)$ times the same mass flow.

Hence two problems arise. One is that of investigating the distribution of flux of additional longitudinal momentum, or (what is the same, when divided by q_∞) that of flux of wave energy, and verifying that its integral over a downstream planè is equal to the drag. The other is that of investigating the distribution of excess mass flow, and verifying that its integral over a downstream plane is zero. This is of significance from the point of view of energy because of the work done by the atmospheric pressure when the density changes.

The latter problem is here dealt with first, as being more directly analogous to the problem of downwash distribution which has just been treated. As in that problem, first consider the distribution for arbitrary $x > c$, but with squares of all disturbances neglected. Then the excess mass flow across the line $x = \text{const}$ where it intersects the upper simple wave region is determined (as Eq. 4-24 was determined) to be

$$\rho_\infty q_\infty y_t \tag{4-35}$$

while that across the part where it intersects the lower simple wave region is equal in magnitude but opposite in sign. There is no contribution, to the first order, from behind the rear shock. Thus, for a lifting airfoil (for which Eq. 4-35 is negative), the increased volume above the dividing streamline downstream of the airfoil is, to the first order, never compensated. The deficiency in mass flow which it produces is concentrated, to the first order, in the wave stretching out diagonally downstream between the shocks. This wave contains the same deficiency of mass flow at any station downstream because its width increases as its amplitude decays. The decreased volume below the dividing streamline similarly causes an equal excess of mass flow propagated diagonally downstream in the lower simple wave. There is a positive downstream flux of total energy $-p_\infty y_t$ in this latter wave, due to the work done by the pressure of the incident stream in compressing the air, and a negative flux equal in magnitude in the upper wave. The total flux of total energy has no first order term, but the continual further separation by the airfoil's motion of regions with excess and deficient total energy is doubtless associated with the lift on the airfoil.

Passing to the second order distribution of excess mass flow we confine ourselves (as we did with the downwash) to its determination for very large x, for which all disturbances have become much weaker than their initial values. Then the calculation of the contribution from the simple waves is straightforward, giving (cf. Eq. 4-26)

$$\rho_\infty q_\infty \left[y_t + K \int_0^c y_u'^2(x)dx \right] + \rho_\infty q_\infty \left[-y_t + K \int_0^c y_l'^2(x)dx \right]$$

$$= \rho_\infty q_\infty (M_\infty^2 - 1)^{\frac{1}{2}} K \int_0^c \left[y_u'^2(x) + y_l'^2(x) \right] dx \quad (4\text{-}36)$$

where the left-hand terms refer respectively to the upper and lower simple wave. Thus, to the second order, the excess mass flow in the simple wave regions adds up to a positive nonzero quantity. Again the question arises whether this is compensated by reduced mass flow behind the rear shock, or whether the Friedrichs theory is wrong to the second order.

Now behind the rear shock, variations of entropy are of the same order as other disturbances, and must be considered in calculating ρu. Using that, by the equation of state and Bernoulli's equation,

$$\frac{d\rho}{\rho} = \frac{1}{\gamma}\left(\frac{dp}{p} - \frac{ds}{c_v}\right), \qquad q dq = -dh = -\,T ds - \frac{1}{\rho} dp \qquad (4\text{-}37)$$

we obtain that, neglecting the squares of disturbances,

$$\rho u - \rho_\infty q_\infty = \frac{M_\infty^2 - 1}{q_\infty}(p - p_\infty) - \rho_\infty q_\infty \frac{s}{c_p}\left[1 + \frac{1}{(\gamma - 1)M_\infty^2}\right] \quad (4\text{-}38)$$

Eq. 4-38 shows how additional entropy reduces the mass flow. The excess mass flow behind the rear shocks consists of three parts, one from the pressure wave transmitted through each shock, and one from the increased entropy of the fluid. The former two are calculated as was Eq. 4-28, giving a value

$$-\rho_\infty q_\infty \frac{3D}{2A}\int_0^c [y_u'^2(x) + y_l'^2(x)]dx \qquad (4\text{-}39)$$

for their sum. The latter is

$$-\rho_\infty q_\infty \left[1 + \frac{1}{(\gamma - 1)M_\infty^2}\right]\int_{-\infty}^{\infty} \frac{s}{c_p}\, dy \qquad (4\text{-}40)$$

for some $x > c$. But the specific entropy s is conserved along streamlines, and therefore the integral (4-40) may be taken along each of the four shocks and added up, using Eq. 3-20 to relate the entropy gain to the deflection at a shock. Then the contributions from the two upper shocks are

$$-\rho_\infty q_\infty \left[1 + \frac{1}{(\gamma - 1)M_\infty^2}\right] \int \frac{\gamma^2 - 1}{12} \frac{M_\infty^6}{(M_\infty^2 - 1)^{3/2}} (\theta_f^3 - \theta_r^3) \frac{dx}{(M_\infty^2 - 1)^{1/2}} \quad (4\text{-}41)$$

This is of the same form as Eq. 4-27 and is similarly evaluated, giving finally for the whole excess mass flow behind the rear shocks

$$-\rho_\infty q_\infty \left\{\frac{(\gamma + 1)M_\infty^4}{4A(M_\infty^2 - 1)^{3/2}} [1 + (\gamma - 1)M_\infty^2] + \frac{3D}{2A}\right\} \int_0^c [y_u'^2(x) + y_l'^2(x)]dx \quad (4\text{-}42)$$

This negative term exactly balances the positive one (4-36) since a simple calculation shows that the coefficient in braces is precisely $(M_\infty^2 - 1)^{1/2}K$.

Thus the excess mass flow in the simple waves is principally compensated by deficient mass flow due to the gain in entropy of the fluid. The pressure wave in the region behind the rear shock may fall on either side of the balance, according to the Mach number of the incident stream. For moderate Mach numbers ($1.25 < M_\infty < 2.54$ for small disturbances, when $D < 0$), the pressure wave adds to the excess mass flow; for others it detracts from it.

We pass next to the two identical problems of determining the distributions of flux of additional longitudinal momentum and flux of wave energy and relating these to drag in accordance with Eq. 4-31 and 4-33. This is done to the second order, neglecting cubes of all disturbances, since no first order terms arise. Again, it is particularly easy to do it so far downstream that the squares of the local disturbances may be neglected. The form of the integrand in Eq. 4-33 then makes it especially clear that the only term involving the first power of the disturbances is the Ts term due to specific entropy increment in the specific internal energy e (since $de = Tds - pd\rho/\rho^2$). Hence

$$\mathfrak{D} = \int_{-\infty}^{\infty} \rho_\infty T s \, dy \quad (4\text{-}43)$$

which is evaluated to the second order, as was Eq. 4-40, as an integral taken along all four shocks,

$$\rho_\infty q_\infty^2 \frac{(\gamma + 1)M_\infty^4}{12(M_\infty^2 - 1)^2} \int (\theta_f^3 - \theta_r^3)dx = \frac{\rho_\infty q_\infty^2}{(M_\infty^2 - 1)^{1/2}} \int_0^c [y_u'^2(x) + y_l'^2(x)]dx \quad (4\text{-}44)$$

agreeing with the Ackeret value.

For general values of $x > c$, the integral (4-31) or (4-33) consists of a portion (4-43), which however is less than its value for $x \to \infty$ because it only comprises the entropy increment at the part of the shock ahead of the plane normal to the stream which is considered, and also a portion due to isentropic disturbances in the simple waves. Actually an easy calculation shows that in any simple wave the integrand in Eq. 4-31 or 4-33 is $\rho_\infty q_\infty^2 \theta^2$, to the second order. Integrated with respect to y between the shocks, this yields precisely that part of the integral (4-43) due to the parts of the shocks downstream of the plane considered. Thus the total remains the same, and equal to the drag.

The expression (4-43) is not absolutely exact since the squares of entropy increments have been neglected. But an exact expression is obtainable. For in the limit as $x \to \infty$ all disturbances in p and θ vanish, and the integral of their square vanishes. Hence the drag is exactly equal to the integral (4-31) or (4-33), with the integrand regarded as a function of the specific entropy, and with the pressure and flow direction at their initial values. Since entropy is constant along a streamline, the expression $\rho u dy$ at constant x in either integral may be replaced by $\rho_\infty q_\infty dy$ along the front shock. For the rear shock it is similarly the element of mass flow across the shock, i.e. it is $d\psi$ if ψ is the stream function. Hence Eq. 4-31 may be written as

$$\mathfrak{D} = \int \{q_\infty - [q_\infty^2 + 2h(p_\infty, 0) - 2h(p_\infty, s)]^{\frac{1}{2}}\} d\psi \qquad (4\text{-}45)$$

taken along the rear shocks, or along any curve stretching right across the stream behind them. Here the velocity u has been expressed in terms of the specific enthalpy $h(p, s)$ by Bernoulli's equation, with $p = p_\infty$ and $\theta = 0$, but with the specific entropy s given its value behind the rear shock. Eq. 4-43 above is merely the first term when Eq. 4-45 is expanded in powers of s.

The analysis of drag in terms of wave energy is convenient because wave energy is proportional to amplitude squared, and hence for example an N wave, for which ultimately wavelength times amplitude is constant, has wave energy tending to zero. However, the analysis might sometimes be misleading; it depends essentially on the fact that the total space available to the fluid remains unchanged as it flows past the airfoil, for only then will the total work done by the pressure of the incident stream be zero. For example, in the flow past a "half-body" shape, which stretches with finite thickness to infinity, Eq. 4-33 would be wrong for this reason. Also Eq. 4-30 and 4-31 would be wrong because the momentum flux and normal forces for the upstream and downstream planes, which are being subtracted, have been taken within the same integral, while actually the limits of integration are different for each. Only Eq. 4-32 remains true in this problem, since on the one hand the total energy flux is considered therein, and on the other hand there is no such flux across the upstream plane. These considerations show that there is some fundamental importance in Eq. 4-32, which relates drag to total energy flux. Therefore, we describe the distribution of the integrand, which is at once deducible by adding wave energy flux and $(p_\infty/\rho_\infty q_\infty)$ times the excess mass flow. The total energy flux in the simple waves, far downstream, is

$$p_\infty(M_\infty^2 - 1)^{\frac{1}{2}}K \int_0^c [y_u'^2(x) + y_l'^2(x)]dx \tag{4-46}$$

(At positive angle of attack the contribution to Eq. 4-46 from the upper simple wave is less than that for the lower one; the contributions may even be negative and positive respectively.) On the other hand, the total energy flux behind the rear shock is equal to the drag (4-44) minus the term (4-46). It is interesting to observe that a proportion

$$\frac{p_\infty(M_\infty^2 - 1)^{\frac{1}{2}}K}{\rho_\infty q_\infty^2(M_\infty^2 - 1)^{-\frac{1}{2}}} = 1 - \left(\frac{\gamma + 1}{8\gamma}\right)\left(\frac{3M_\infty^2 - 4}{M_\infty^2 - 1}\right) \tag{4-47}$$

of the drag goes to sending energy to infinity diagonally along the simple waves. For $M_\infty \to \infty$ this is only $\frac{5}{14}$ of the total, a fraction already calculated by the author [30] in his discussion of the analogous problems in one-dimensional unsteady flow, from which the above account of momentum and energy balance has been adapted. But as M_∞ decreases the fraction (4-47) increases, until for $M_\infty < \sqrt{\frac{4}{3}} = 1.154$ it exceeds unity, and the energy flux behind the rear shock is negative because that due to the rear pressure wave is negative and exceeds in magnitude the essentially positive contribution from the increased entropy. This latter contribution is a fraction

$$\frac{M_\infty^2 - 1}{\gamma M_\infty^2} \tag{4-48}$$

of the whole, which is $\frac{5}{7}$ for $M_\infty \to \infty$ but much less for smaller M_∞.

Of course, actually, the pressure wave is also propagated diagonally to infinity, so that the proportion of total energy flux so propagated is not Eq. 4-47 but is 1 minus Eq. 4-48. With regard to this pressure wave there is one additional complication which still has to be mentioned. This is that it overtakes the rear shocks. For Eq. 4-12 of the rear shocks makes it clear that every Mach line, originating however far behind it, must at last intersect it when $x^{\frac{1}{2}}$ has become sufficiently large. Actually the rear pressure waves are very broad, but as the rear shocks are followed backwards they must gradually be intersected by more and more of the pressure waves (see Fig. E,4b). Now these waves are already of the third order, but so is the reflection coefficient for reflection at the shock. Hence their reflection at the shock constitutes a negligible wave. But their energy cannot be transmitted through the shock since the Mach lines ahead of the shock both point upstream. Hence their energy goes simply into *changing the shock path from that given by the Friedrichs theory*, making the shock decay less or more rapidly according as the said energy is positive or negative (i.e. according as $1.25 < M_\infty < 2.54$ or not). Finally, then, all the energy, which was shed by the N wave as a reflected and transmitted pressure wave, is reabsorbed in the N wave by the overtaking of the rear shocks by acoustic waves traveling behind them. When this has happened the proportion of total energy flux in the N wave is simply 1 minus Eq. 4-48.

From this knowledge of the ultimate total energy flux in the N wave, after the whole broad tail pressure wave has caught up with the rear shock, the final asymptotic position and strength of the rear shock can be calculated. It is exactly as in Eq. 4-10 and 4-12 but with K replaced by

$$\frac{1 + (\gamma - 1)M_\infty^2}{M_\infty^2 - 1} \tag{4-49}$$

Thus the shape of the rear shock gradually alters slightly, so that in its asymptotic form the coefficient of

$$\int_0^c y_u'^2(x)dx$$

changes from K to the expression (4-49), as the rear pressure wave overtakes it. This is the one point in which our critical analysis of the Friedrichs theory has found it deficient. It gives the front shock correctly to a second approximation, but after a long time begins to give the rear shock correctly only to a first approximation, although by a modified theory we can determine the total departure from the theoretical form to the second approximation. Of course the check relating lift and downwash is not affected, since the downwash in the tail pressure wave is absorbed in the N wave as the process goes on.

E,5. Supersonic Two-dimensional Airfoil Theory: Influence of a Blunt Leading Edge. In this article some attempt is made to extend the theory of Art. 3 and 4 to the flow past airfoils with blunt leading edges. It is important to notice that this includes all real airfoil sections. It is not possible to make a leading edge of true wedge shape. With wedge angles such as the theory indicates are desirable, careful construction will produce a leading edge "thickness" (i.e. perpendicular distance between nearest parts of surface which are truly inclined) at best of the order of 5 microns. Further, the edge so produced is very fragile indeed (note that, of course, *cutting* edges are not designed of wedge section) and, after a few minutes in a supersonic wind tunnel, it may have deteriorated so that the effective leading edge thickness is roughly of the order of 20 microns. If this happens, the spanwise distribution of leading edge thickness becomes very irregular, and therefore it is probably wise to aim at a value not less than 20 microns in the original design. Of course in many low-precision experiments, with so-called sharp leading edges, the leading edge thickness is 100 microns or more. But it may still be a very small fraction of the airfoil thickness. Thus considerable interest is attached to predictions of how a *slight* blunting of the leading edge will affect the flow, when the conclusions of this article constitute a small correction to the theories of Art. 3 and 4.

In the steady flow of a uniform supersonic stream past any airfoil with a blunt leading edge, the front shock is detached, and there is a region of subsonic flow between it and the leading edge. The theoretical

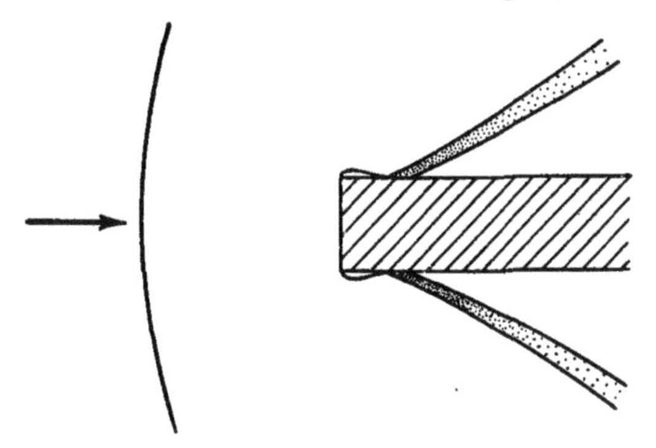

Fig. E,5a. Boundary layer separation in supersonic flow past a square-nosed plate.

prediction of the complete transonic flow field is a very complicated matter (Sec. H). Here, only the region of supersonic flow is discussed theoretically; the influence on it of the subsonic region is taken from experimental data.

The data used are from experiments at the NPL by Holder and Chinneck [*63*] on symmetrical supersonic flow past flat plates with blunt leading edges of elliptical shape. The axis ratio of the ellipses varied from zero (giving a square nose) through unity (semicircular nose) to 8. It will be explained shortly why, although in each case the body shape behind the blunt leading edge was flat, the data can nevertheless be used to predict flows around more general airfoil shapes.

It must be stated at once that, especially for the square-nosed plate, boundary layer effects are observed near the leading edge (where, for the normal development of the Blasius boundary layer, they would be negligible). Thus the flow pattern cannot be completely independent of Reynolds number. The precise effect observed is that the boundary layer appears to separate and become reattached. (This separation doubtless occurs because, if it did not, the flow would overexpand around the shoulder of the plate so much that a large positive pressure gradient must follow.) Where reattachment takes place, the main stream turns rapidly through an angle concave to the flow, producing an oblique shock. Fig. E,5a, which illustrates this behavior for a square-nosed plate, is a fairly close copy of a photograph taken with "micro-schlieren" apparatus by the authors cited above.[1] (The large apparent thickness of the second shock is of course due to the fact that the dead air region is not very accurately two-dimensional.)

However, for the plates whose leading edge had a more "streamline" shape, say semicircular, or semielliptical with axis ratio exceeding 1, the phenomenon just described was much less marked. Only the shock could be observed, not the separation region which produces it. Shadow photographs indicated that this subsidiary shock is very weak in comparison

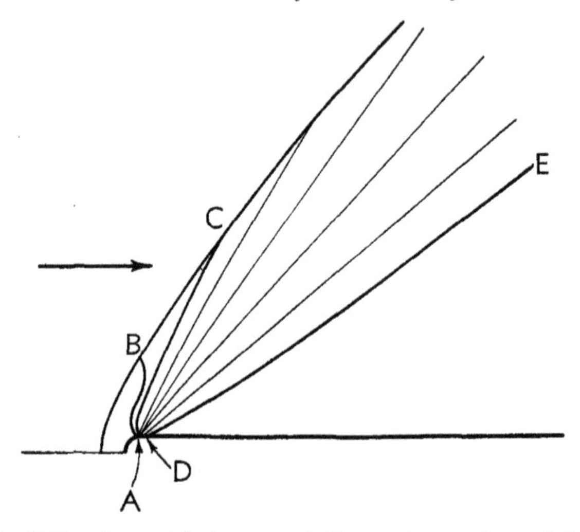

Fig. E,5b. Symmetrical supersonic flow past a round-nosed plate.

with the main front shock. The effective deformation of the surface, produced by the dead air region between it and the separated boundary layer, is apparently unimportant. That is to say, the separation and resulting shock, while of intrinsic interest, appear to have little influence on the rest of the flow. Hence the effect of the boundary layer will be neglected in what follows, except in occasional critical remarks. This will be justified provided that the shoulders at the nose are not too sharp (as in the square-nosed plate). Some attempt to make them more streamlined is clearly of practical importance, since the extra shock (if significantly strong) and its associated entropy change must be connected, as Art. 4 shows, with an increase in drag; and it will be seen below that the drag associated with the main front shock is also reduced by streamlining the leading edge, given its thickness.

Ignoring the boundary layer effects the flow pattern appears to be as

[1] However, Mair and Bardsley [*64*] present evidence that such separation may not occur when the Reynolds number based on leading edge thickness is less than about 1,000.

sketched in Fig. E,5b, which shows only the upper half of the symmetric flow pattern. The shock shape shown is that observed at $M_\infty = 1.8$ for a semicircular leading edge. The sonic line AB is drawn in conjecturally from study of the schlieren photographs. The point B is, however, known for certain from the shock inclination producing sonic conditions behind it. The first Mach line emanating from the surface is AC. Other Mach lines behind it form a fan ending with the final Mach line DE (in reality, as remarked above, a weak shock). Little disturbance is visible behind DE on schlieren photographs, but surface pressure measurements show that slight variations subsist for a considerable length beyond D. The precise nature of these varies considerably with nose shape.

Now it is clear that the flow to the left of AC cannot be influenced by changes in the shape of the boundary downstream of A, for the effect of these would be propagated along Mach lines to the right of AC, which will not cross AC if the boundary curvature is convex to the stream. Conversely, the supersonic flow to the right of AC is determined by the conditions on AC itself and by the shape of the boundary downstream of A. If conditions on AC were uniform, and AC were a straight line, then the flow to the right of AC would be a simple wave, determined by the boundary shape. As it is, the nonuniformity in AC, both in slope and in the distribution of flow quantities along it, causes a certain amount of propagation along Mach lines pointing in toward the body.

The study of the flow is made easier if we use the result of Art. 4 that, in spite of the higher specific entropy of the fluid near the surface, which has passed through a normal shock, the variations of pressure p and flow direction θ can be treated to good approximation as if the specific entropy were uniformly that of the incident stream. Then the pressure wave propagated in toward the airfoil by reason of the nonuniformity on AC may be regarded, together with its reflection at the surface, as simply superimposed on the simple wave which would be produced by the part of the surface downstream of A in continuous shock-free deformation of the incident stream. (The analogy is very strong between this superposed pressure wave and the wave which was shown in Art. 4 to be reflected from an attached shock.)

To a good enough approximation it may be assumed that the effect of this additional pressure wave on the surface pressures (which is double the pressure distribution in the wave, owing to the reflection at the surface) depends only on the nose shape (i.e. we ignore its interaction with the simple wave). Hence experiments, such as were done at the NPL, at three Mach numbers and for six nose shapes, with a perfectly *flat* surface to the right of D, can be used to determine the pressure wave in the general case, at least as far as its influence on surface pressures to the right of D is concerned. For in the simple wave there would, for the NPL shapes, be no departure from the undisturbed pressure downstream of D, and therefore the said *departures* constitute the whole of what must, in other problems, be added to the pressure distribution predicted by simple wave theory.

In Fig. E,5c this excess pressure distribution downstream of D is shown, as measured at the NPL for $M_\infty = 1.6$ and elliptic noses of axis ratios 0 (square nose), $\frac{1}{2}$, 1, 2, 4, and 8. The distributions, and their

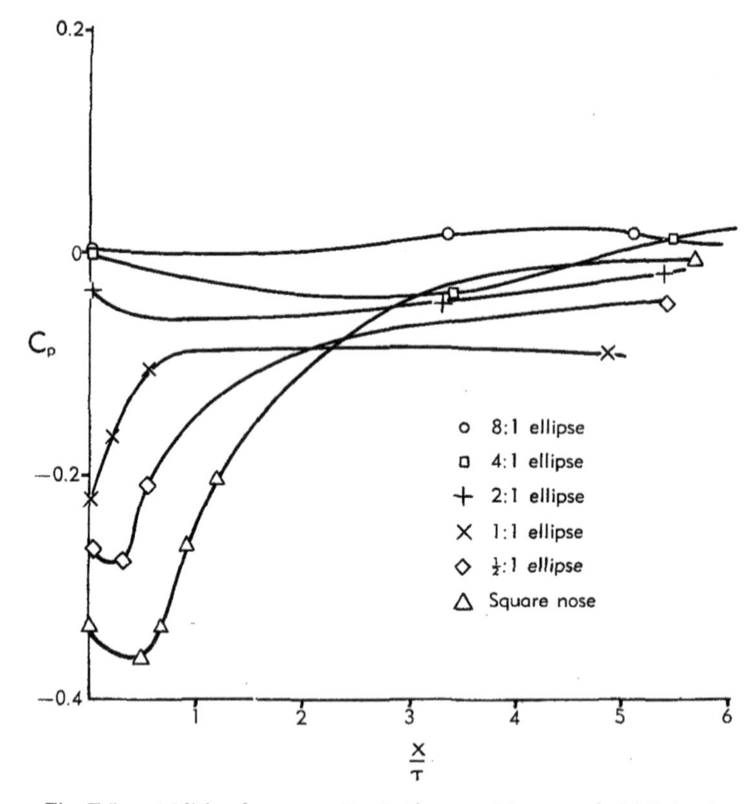

Fig. E,5c. Additional pressure distribution on a blunt-nosed airfoil due to departure of the flow from simple wave form, for different nose shapes.

dependence on nose axis ratio, were closely similar for $M_\infty = 1.42$ and $M_\infty = 1.79$. It is worth remarking that these Mach numbers are *all* in the range where the coefficient D of Art. 3 and 4 is negative, so that any part of the wave arising simply from reflection of a simple wave at the front shock would be *compressive* ($C_p > 0$).

It must next be inquired what effect the blunted nose is likely to have on the aerodynamic forces, and particularly on the drag. Again it is clear that the forces due to the nose will, in the main, be additive to those due to the simple wave flow behind it, the principal contribution being from the region upstream of D in Fig. E,5b. The importance of the additional pressure wave behind D, which is plotted in Fig. E,5c, will, however, depend on how much the airfoil continues to thicken in this region, and so will not be additive. The drag due to the pressures on the nose itself is deducible at once from the NPL experiments, and is plotted against M_∞ for the six different nose shapes in Fig. E,5d. The ordinate is

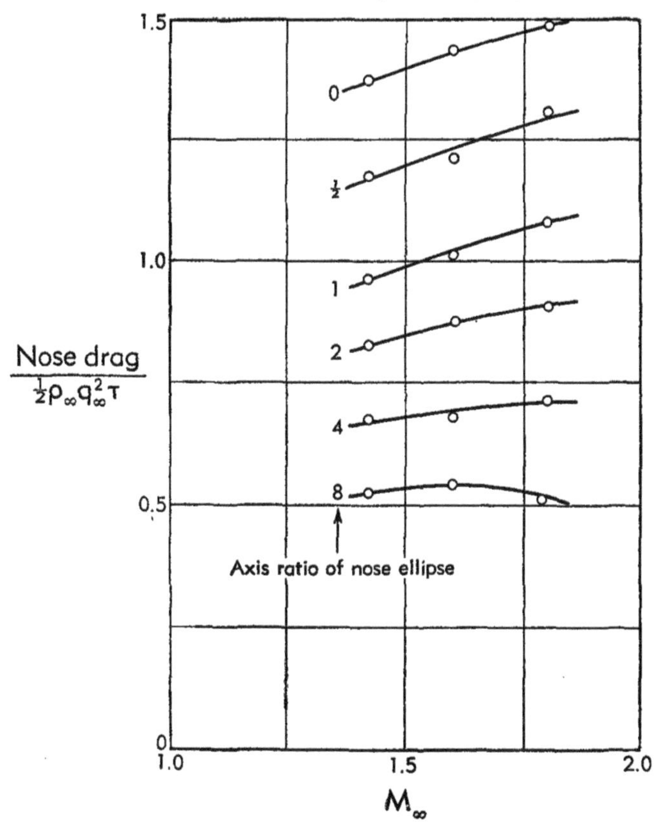

Fig. E,5d. Nose drag at Mach number M_∞, for different nose shapes with thickness τ.

drag/$(\frac{1}{2}\rho_\infty q_\infty^2 \tau)$, where τ is the thickness of the leading edge. For an airfoil the contribution of the nose drag to C_D will therefore be (τ/c) times the ordinate in Fig. E,5d.

It is seen that this quantity varies critically with the axis ratio of the nose ellipse, and that, for values of τ exceeding $0.002c$, significant drag reduction can be obtained by making the axis ratio large rather than using, say, a square nose. Some understanding of the rapid rise of drag as the nose axis ratio decreases can be obtained by plotting pressure coefficient against flow direction over the various noses, for one Mach number (actually 1.79), on a single figure, Fig. E,5e. It is seen that for axis ratios 2, 4, and 8 the points are scattered about Ackeret's straight line $C_p = C_1\theta$, even in the subsonic region, though for smaller axis ratios they begin to fall consistently below it. At least there is some indication that pressure is more closely related to flow direction than to various characteristic distances (plotted against which, the pressure graphs are widely separated in the different cases). Hence for given thickness τ, the drag would be expected to be least if the flow direction *averaged with*

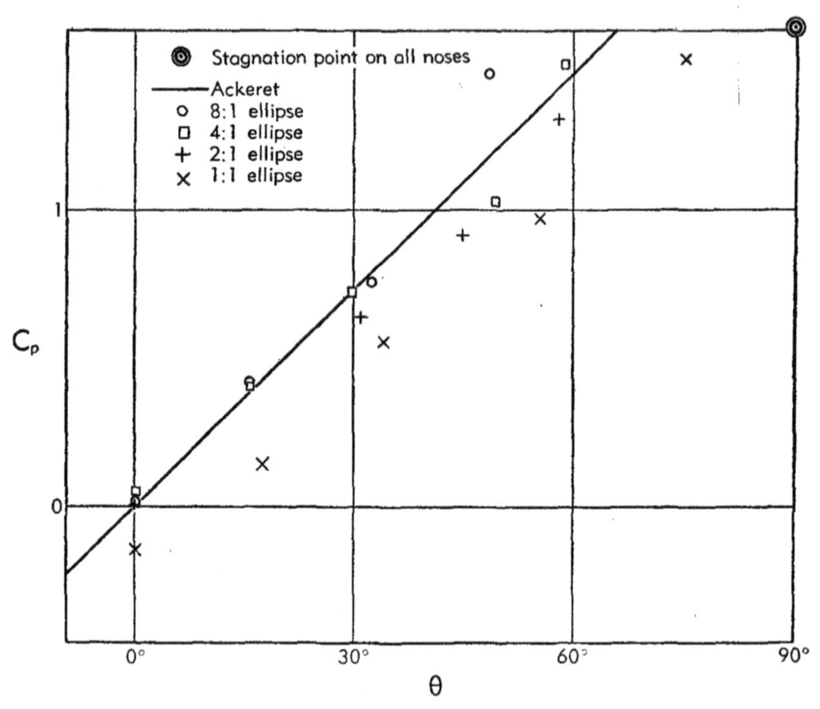

Fig. E,5e. Pressure coefficient on a blunt nose, plotted against flow direction.

respect to thickness is least. This is so for the elliptic noses of larger axis ratio, for which the rise in θ from 0° to 45°, for example, occurs in a larger region (viewed head-on) than the subsequent rise to 90°.

It is significant that the drag coefficient rises with Mach number, in the interval considered, for the smaller axis ratios. For very large axis ratios (nearly sharp noses) it must *fall* with Mach number, as Ackeret's formula indicates. But for very blunt noses, where most of the drag is probably associated with entropy change at the nearly normal part of the shock, the drag coefficient would be expected to rise, since the strength of this part of the shock increases with Mach number, and with it the entropy change increases rapidly.

To obtain the whole pressure drag at zero lift on a blunt-nosed airfoil, one must add the nose drag, just discussed, to the drag predicted by Busemann's approximate simple wave theory applied over the rest of the airfoil, and if necessary include also the drag due to the additional pressure distribution illustrated in Fig. E,5c. The latter is obtained by multiplying the said distribution of C_p by the flow direction θ, integrating along the chord, and multiplying by $2/c$. The resulting contribution, when significant, is negative, and due to the dip at the beginning of each

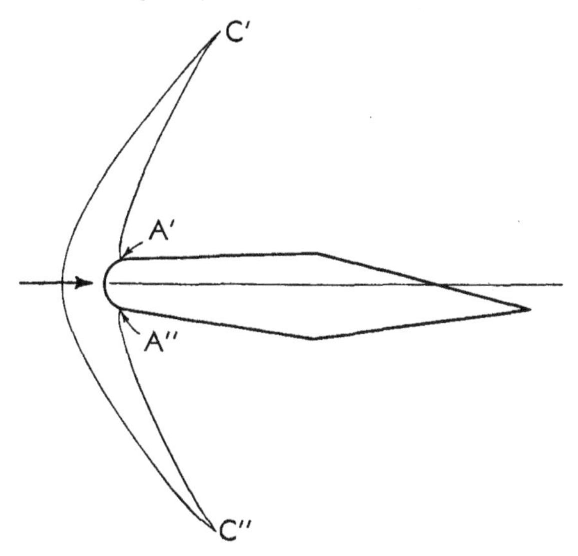

Fig. E,5f. Supersonic flow past a round-nosed airfoil at angle of attack.

of the curves with axis ratio $\leqq 1$ in Fig. E,5c. On the other hand, this will be partly balanced by a smaller negative pressure coefficient acting over a slightly larger frontal area at the rear, and probably for most practical purposes it may be ignored. Then effectively *the drag is increased by the blunting of the leading edge exactly as it would be if the pressure coefficient over the leading edge thickness were raised from its value for a perfectly sharp edge to that given by the ordinates in* Fig. E,5d.

There have not been such systematic measurements of the effect of leading edge thickness on lift, and on the extra drag due to angle of attack, but it is generally believed that these effects are not very important. There is a theoretical argument indicating this for the case of a semicircular leading edge. For then the flow pattern at angle of attack must be as in Fig. E,5f, where a very large leading edge thickness has been chosen for the sake of clarity. The flow to the left of $A'C'$ and $A''C''$ is identical with that to the left of AC, and of its reflection in the axis of symmetry, in Fig. E,5b. For the nose shape is invariant under rotation, and the fact that the surfaces downstream of A and A' are not identical cannot make a difference upstream of AC and $A'C'$. Hence no resultant lift, or additional drag, arises from this region, and therefore it is clear (from Eq. 3-22 and 3-23, which hold over the rest of the surface) that, if a factor of the form $1 + \tau/c$ is negligible, both these quantities are unchanged by making the leading edge blunt. Thus the main effect of blunting the leading edge is to increase the drag at zero lift as already analyzed in Fig. E,5e.

Finally, some brief mention must be made of the shape of the front shock. Its shape in the subsonic region is discussed elsewhere (Sec. H) both theoretically and experimentally, but it is desirable here to discuss its behavior farther out and the connection of this behavior with the theory of Art. 4.

It is obvious that the Friedrichs theory (Art. 4) applies to the shock above C. Indeed the streamline through C may be regarded as a thin sharp-nosed airfoil producing a shock exactly as in Art. 4, but more fundamentally one can regard the shock shape above C as deducible from Mach lines in Friedrichs' form (4-1), where, however, $Y(\theta)$ is related to the surface shape less directly than when the shock is attached. For the blunt flat plate, for which the Mach lines concur to good approximation in a region around AD, the asymptotic formulas may be expected to hold fairly well already at, say, ten leading edge thicknesses from the body, by the author's geometrical argument (Art. 4) leading to the parabolic asymptotic form. For example, the strength $(\Delta p/p)$ of the shock, multiplied by the square root of distance $\sqrt{y}$ from the axis of symmetry, should be nearly uniform, but would not be expected to equal the value of $2\gamma\sqrt{[y_p/\gamma + 1]}$ given by the first approximation of Art. 4, since the relation of $Y(\theta)$ to body shape is different. Indeed, with y_p replaced by $\frac{1}{2}\tau$, this approximation to $(\Delta p/p)\sqrt{y}$ for large y is $1.3\sqrt{\tau}$, while the observed values are between two and three times as much. To within an error of ± 0.2, the values of $(\Delta p/p)\sqrt{y/\tau}$ for values of y/τ exceeding 10, for both $M_\infty = 1.6$ and 1.79, are as in the following table. We see that a round figure for blunt noses in general is 3.

Axis ratio of nose	0	$\frac{1}{2}$	1	2	4	8
$\lim_{y\to\infty} \Delta p/p\sqrt{y/\tau}$	3.4	3.2	3.0	2.8	2.6	2.4

But when the surface behind the blunted nose makes a positive angle with the incident stream, then large curvature of the surface ceases at some point D' ahead of D. This earlier flattening out is signaled along the Mach line through D', and the shock is altered from that described in the last paragraph after encountering this Mach line. It decays more slowly after this encounter, and indeed to any straight portion of surface behind D' corresponds a straight portion of shock of uniform strength.

On the other hand, if the angle of attack exceeds the inclination of the upper surface behind the blunt leading edge, then the shock is precisely as described two paragraphs above. For the Mach lines from the upper surface behind the leading edge then interact only with the rear shock. For such a flow, if the leading edge had been sharp, the flow over the upper surface would have been headed by a Prandtl-Meyer expansion. But in reality this is preceded by a shock whose shape must be exactly as in the NPL experiments on blunt-nosed plates. In particular its strength is given as

$$\frac{\Delta p}{p} \cong 3\sqrt{\frac{\tau}{y}} \tag{5-1}$$

when $y/\tau > 10$.

Bardsley [*65*] shows that even for the smallest leading edge thicknesses obtainable, this simple consideration is sufficient to account for the shocks which have always been observed to precede the Prandtl-Meyer expansion at the leading edge of a supersonic airfoil at high angle of attack. For example Eq. 5-1 predicts that, for a leading edge thickness of 20 microns, the shock strength at a distance 5 cm from the leading edge measured perpendicular to the stream, should be 0.06. (Shocks of such a strength are easily visible on schlieren photographs.) This value is, within the experimental error, the strength calculated by Bardsley from the measured shock inclination under such conditions. Bardsley also gives calculations leading to the rejection of the alternative theory that the shock in question is produced by boundary layer growth.

To sum up the present article, it has been demonstrated very briefly that, although the theories of Art. 3 and 4 do not apply directly to blunt-nosed airfoils, nevertheless suitable semiempirical modifications of these theories can in principle be constructed to give as much information about every aspect of the flow as we have in the simpler problems of sharp-nosed airfoils. But the carrying out of this program in full detail may perhaps have to wait until an even more systematic series of experiments than those described above has been performed.

E,6. Supersonic Projectile Theory: Complete Flow Pattern. In Art. 6 and 7 we seek information, of the kind obtained for airfoil flow in Art. 3, 4, and 5, in the analogous problem of axisymmetric flow past projectiles, and also in some problems where flow is not strictly axisymmetric but the projectile surface everywhere makes a fairly small angle with the stream. Even the axisymmetric flow problem is a more complicated one than that of airfoil flow, largely because the solution on small perturbation theory (D,15–D,18) is more complicated, which in turn is due to the famous fact that a cylindrical pulse of sound, unlike a plane or spherical pulse, leaves a "tail" behind it (cf. C,12).

In consequence, fewer strictly logical inferences are possible. But this is partly compensated by the possibility of arguments by analogy from the two-dimensional case. For example, it will be assumed that, at least for sharp-nosed projectiles with attached shocks and a completely supersonic flow pattern, the flow is given to excellent approximation by the potential flow, at the entropy of the undisturbed stream, which satisfies the boundary condition at the surface of the projectile—always provided that this potential flow is cut off by the front shock (which is to be determined by the conditions governing shocks with uniform flow ahead of them) to avoid the appearance of a region in which there are three values for each physical quantity, and that similarly, where necessary, a rear shock or shocks appear. The assumption here is by analogy with the "simple wave theory" of Art. 3, whose accuracy in predicting surface pressures was amply demonstrated in Fig. E,3f, and whose accuracy in predicting shock positions was carefully checked in Art. 4. The errors, being due to those differences between the behavior of shocks and of acoustic discontinuities which were collected in the last sentence of point (7) of Art. 1, should be no greater in the present problem since the shocks, which still have supersonic flow on both sides of them, can be no stronger than before.

Such analogies must of course be used intelligently. There is no reason to suppose that any of the flow quantities are constant along Mach lines, or that these are straight (the first point is not even true on the small perturbation theory). For these Mach lines are meridian sections of stationary sound waves of conical, or nearly conical, shape. There is no reason why they should remain exact conical waves, which have not the self-preserving property of plane waves. Again, it should not be assumed that the pressure at a point depends only on the flow direction, or that the flow is uniform behind the rear shock, even on this simplified theory, or indeed even on linear theory, owing to the "tail" property mentioned above.

Thus, although it is at least true that the essential problem is the determination of a potential flow, it is clear from the points in the last paragraph that a method like Busemann's (Art. 3) for proceeding to the second approximation to the surface pressures will not be available in axisymmetric supersonic flow. Therefore a straightforward approach by successive approximations is necessary, as for the subsonic flow problems of Art. 2, and especially as in the Hantzsche-Wendt method. But such a method is much more laborious in axisymmetric problems. Actually, as Broderick [*34*] found, and as will be reported briefly in Art. 7, it becomes simpler when only the pressures close to the projectile surface are sought, but it is still almost too complicated to be practically useful, although the situation has recently been improved by Van Dyke [*54*]. Further, even if this second approximation could be found away from the body surface, it would give one no information about the position of the shocks, which are among the most interesting features of the flow field. To see this, e.g. for the front shock in axisymmetric flow, it is sufficient to observe that the equation for the second approximation to the disturbance potential (which is derived from Eq. 2-14) has the same characteristic curves as the equation for the first approximation, namely, the straight lines whose equations in cylindrical polar coordinates are $x \pm (M_\infty^2 - 1)^{\frac{1}{2}}r = \text{const}$. Hence on the second approximation the flow is still uniform ahead of the downstream-pointing characteristic, given by $x - (M_\infty^2 - 1)^{\frac{1}{2}}r = \text{const}$, through the nose of the projectile, whereas in reality the front shock must lie ahead of this line. Actually, as is hardly surprising after this consideration, the process of successive approximation diverges near any shock, as foreshadowed in Art. 1. A similar difficulty arises at points whose distance r from the axis of the projectile is large. Owing to the continual separation, as r increases, of the approximate (straight) Mach lines from the true ones, the second approximation may be as bad as the first for large r; and indeed for sufficiently large r the process again diverges.

However, there is a quite different approach by which all this latter information (concerning the approximate strength and position of shocks, and the flow at large distances from the axis, which the second approximation as well as the first fails to give) may be obtained by a fairly direct and physically natural device from the form of the first approximation alone. This approach is due to Whitham, who has improved on his previous work [*32*] in a later paper [*66*]. The theory gives all the information analogous to that resulting from Friedrichs' theory of two-dimensional flow (Art. 4), but of course only to a first, not to a second, approximation.

Since the Whitham theory of the complete flow pattern is simpler than the Broderick theory of the surface pressures, it is here given at once, leaving the Broderick theory to Art. 7, and thus reversing the order followed in Art. 3 and 4 for the corresponding two-dimensional problem. This order also reflects the greater practical importance, in the present problem, of a knowledge of the complete flow pattern, which satisfies the assumptions of the theory at *all* distances from the projectile (while the flow past wings, even of large aspect ratio, is not nearly two-dimensional at large distances, as mentioned at the beginning of Art. 4). In fact the conclusions of this article may reasonably be expected to give the correct order of magnitude of the disturbances, at sufficiently large distances, for a projectile of any shape whatever, because the theory at least works in the right number of dimensions. In other words some knowledge of what should be observed at ground level, when any supersonic projectile passes overhead, must emerge.

The Whitham theory. Whitham's theory [*66*] will be described for axisymmetric flow, and afterward we shall show that it holds without modification in more general flows. The cardinal assumption on which it rests is as follows. The small perturbation theory gives a correct first approximation throughout the flow, *provided that the value which it predicts for any physical quantity, at a given distance from the axis on the straight (approximate) Mach line,* $x - (M_\infty^2 - 1)^{\frac{1}{2}}r = \text{const}$, *pointing downstream from a given point on the projectile surface, is interpreted as the value, at that distance from the axis, on the exact Mach line which points downstream from the said point.* In other words the ineffectiveness, near the shocks and at large distances from the axis, of the linear theory is due to its incorrect placing of the Mach lines, not to incorrect prediction of the variation of physical quantities along them.

The following arguments support Whitham's assumption:

1. It is true, word for word, in the two-dimensional flow past airfoils. Applied there it would give the first approximation to the Friedrichs theory, which has been copiously checked in Art. 4.

2. It takes into account the convection of sound, and the variations in its speed (Art. 1), for the features of the problem through which they might be expected to act most effectively, namely for the positions of waves, not the state of the fluid thereon.

3. It includes the statement that the small perturbation theory gives a correct first approximation to the surface pressures, which is supported by strong evidence, some given in VII, B and some in Art. 7 below.

4. The conclusions are not in contradiction with any known facts, and several of them agree with results obtained previously by quite different methods, as is mentioned below in connection with each such conclusion.

The practical use of the assumption depends on the fact that, if the linearized solution is known, then a second approximation to the Mach lines may be deduced. If this second approximation is used, instead of the exact Mach lines mentioned in Whitham's assumption, the approximation should still be reasonable. The spread of the Mach lines at large distances (which causes the decay of the shocks) will be given, not accurately, but to an approximation probably adequate in practice.

To find this second approximation to the Mach lines, let the velocity components parallel and perpendicular to the axis be $q_\infty(1 + u)$ and $q_\infty v$ respectively, so that u and v are nondimensional disturbance velocities. (Then the pressure is given, on small perturbation theory, as $p_\infty - \rho_\infty q_\infty^2(u + \frac{1}{2}v^2)$, where the $\frac{1}{2}v^2$ need only be included near the axis, a point which has been mentioned in Sec. C and D and will be more fully substantiated by the nonlinear theory in Art. 7.) The Mach directions at any point satisfy

$$\frac{dx}{dr} = \cot(\pm\beta + \theta) \tag{6-1}$$

where, as in Art. 3, β is the local Mach angle and θ the direction of the stream, measured in the positive sense from the axis of symmetry. For the Mach direction pointing outward and downstream the plus sign must be taken in Eq. 6-1. For this Mach direction, Eq. 6-1 may be expanded, using Bernoulli's equation, in the form

$$\frac{dx}{dr} - (M_\infty^2 - 1)^{\frac{1}{2}} = \frac{(\gamma + 1)M_\infty^4}{2(M_\infty^2 - 1)^{\frac{1}{2}}} u - M_\infty^2[v + (M_\infty^2 - 1)^{\frac{1}{2}}u] \quad (6\text{-}2)$$

if the squares of disturbances are neglected. (This result is more complicated than the analogous result (4-4) for a simple wave, in which to a first approximation $v + (M_\infty^2 - 1)^{\frac{1}{2}}u = 0$.)

Now the values for u and v given by small perturbation theory (D,16), expressed as functions of r and of the variable $y = x - (M_\infty^2 - 1)^{\frac{1}{2}}r$,

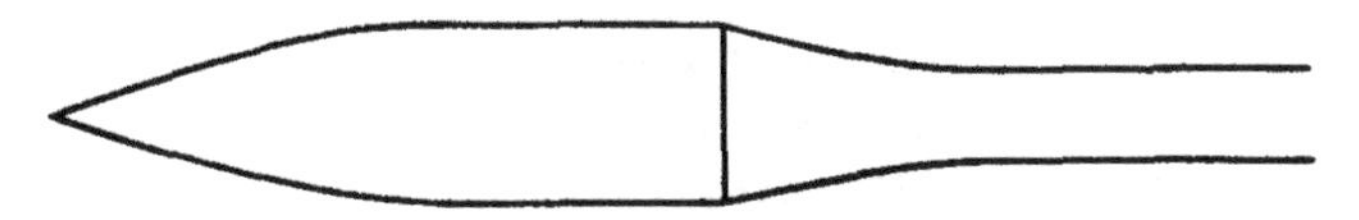

Fig. E,6a. Body-wake combination.

which determines on which straight (approximate) Mach line a point lies, are

$$u = -\int_0^y \frac{f'(t)dt}{(y - t)^{\frac{1}{2}}[y - t + 2(M_\infty^2 - 1)^{\frac{1}{2}}r]^{\frac{1}{2}}}$$

$$v = \frac{1}{r}\int_0^y \frac{[y - t + (M_\infty^2 - 1)^{\frac{1}{2}}r]f'(t)dt}{(y - t)^{\frac{1}{2}}[y - t + 2(M_\infty^2 - 1)^{\frac{1}{2}}r]^{\frac{1}{2}}} \quad (6\text{-}3)$$

The function $f(t)$ in these integrals depends on the body shape, and if the flow behind the body is to be determined the "body shape" should be understood to include also the effective shape of the wake (a sort of "mean" shape, in the sense of turbulence theory). In this sense a typical "body shape" is as illustrated in Fig. E,6a.

For a slender pointed body, the slope and curvature of whose surface vary only gradually, the function $f(t)$ is equal to $S'(t)/2\pi$, where $S(t)$ is the cross-sectional area of the body (or the effective cross-sectional area of the wake) at a distance t (measured along the axis) from the nose. But if, for example, $S'(t)$ (and hence the slope of the surface) is discontinuous anywhere, the function $f(t)$ lags behind, changing to the new value continuously over a distance of a radius or two; this makes a significant difference in Eq. 6-3, as Lighthill [*35*] showed (Art. 7), since in the region of rapid change $f'(t)$ is relatively large. Where the slope or curvature varies rapidly one may correct $f(t)$ suitably by replacing the said rapid variation by a number of small discontinuities of slope, and applying the author's theory to each. Finally, if the body is *not* slender, expressions of the form of Eq. 6-3 may be too inaccurate near the body because the disturbances are not small; but nevertheless at distances from the axis exceeding some suitable multiple of the radius of the projectile, in other words outside a certain circular cylinder enclosing the body, the disturbances will be small enough, and here at least equations of the form of Eq. 6-3 are valid. The function f in these equations may still sometimes be estimated, e.g. from experimental data, using the full small perturbation theory as a partial guide; whenever this function is known the theory below gives the complete flow pattern outside the cylinder mentioned.

Now Whitham's assumption, italicized above, implies that Eq. 6-3 give a valid first approximation (for small perturbations) *everywhere*, provided that in them y stands not for $x - (M_\infty^2 - 1)^{\frac{1}{2}} r$ itself, but for a quantity constant along any downstream-pointing Mach line and equal to $x - (M_\infty^2 - 1)^{\frac{1}{2}} r$ at the surface. Henceforth y will be understood in this sense. *To a close approximation it is where the Mach line, produced, meets the axis* (measured downstream from the nose).

Now we can obtain the relation between x, r, and y to a second approximation (the first being $x - (M_\infty^2 - 1)^{\frac{1}{2}} r = y$) by integrating Eq. 6-2, which holds all along any downstream-pointing characteristic (along which y is constant), starting from the initial value given by $x - (M_\infty^2 - 1)^{\frac{1}{2}} r = y$. This gives

$$x - (M_\infty^2 - 1)^{\frac{1}{2}} r = y + \int_R^r \left\{ \frac{(\gamma + 1) M_\infty^4}{2(M_\infty^2 - 1)^{\frac{1}{2}}} u - M_\infty^2 [v + (M_\infty^2 - 1)^{\frac{1}{2}} u] \right\} dr \quad (6\text{-}4)$$

where the integration is performed at constant y. Here R signifies the radius of the projectile where the Mach line specified by y springs from it, i.e. where $x - (M_\infty^2 - 1)^{\frac{1}{2}} r = y$. (But for nonslender bodies Whitham's assumption can be applied only to the flow outside a certain cylinder enclosing the body, and then R may be taken to be the radius of this cylinder.)

By Eq. 6-3, Eq. 6-4 may be explicitly integrated to give

$$\begin{aligned} x - (M_\infty^2 - 1)^{\frac{1}{2}} r &= y \\ &+ \frac{(\gamma + 1) M_\infty^4}{2(M_\infty^2 - 1)} \int_0^y f'(t) \frac{[y - t + 2(M_\infty^2 - 1)^{\frac{1}{2}} R]^{\frac{1}{2}} - [y - t + 2(M_\infty^2 - 1)^{\frac{1}{2}} r]^{\frac{1}{2}}}{(y - t)^{\frac{1}{2}}} \, dt \\ &- 2M_\infty^2 \int_0^y f'(t) \ln \left\{ \frac{[y - t + 2(M_\infty^2 - 1)^{\frac{1}{2}} r]^{\frac{1}{2}} - (y - t)^{\frac{1}{2}}}{[y - t + 2(M_\infty^2 - 1)^{\frac{1}{2}} R]^{\frac{1}{2}} - (y - t)^{\frac{1}{2}}} \frac{R^{\frac{1}{2}}}{r^{\frac{1}{2}}} \right\} dt \quad (6\text{-}5) \end{aligned}$$

This is the second approximation to the equation of the Mach line specified by y. However, it is never used in the Whitham theory in this very complicated form, but always in the approximate form

$$x - (M_\infty^2 - 1)^{\frac{1}{2}} r = y - kF(y) r^{\frac{1}{2}} \quad (6\text{-}6)$$

where

$$k = 2^{-\frac{1}{2}}(\gamma + 1)M_\infty^4(M_\infty^2 - 1)^{-\frac{3}{4}}, \qquad F(y) = \int_0^y \frac{f'(t)dt}{(y - t)^{\frac{1}{2}}} \tag{6-7}$$

This is obtained from Eq. 6-5 by assuming that $(M_\infty^2 - 1)^{\frac{1}{2}}r$ is large compared with y, so that the form of the characteristic is considered at a point considerably farther downstream from its point of origin on the surface than the said point of origin is from the nose. The justification for making this approximation everywhere is that, where $(M_\infty^2 - 1)^{\frac{1}{2}}r/y$ is not large, the integrals in Eq. 6-5 are unimportant compared with the main term y, which is obvious because the range of integration is y but the integrand is of the same small order of magnitude as the disturbances. Hence these integrals can be approximated as $-kF(y)r^{\frac{1}{2}}$, because this also is unimportant when $(M_\infty^2 - 1)^{\frac{1}{2}}r/y$ is not large. Actually it will appear below that, wherever the small perturbation theory in its ordinary form is inadequate, including all regions near shocks, the condition that $(M_\infty^2 - 1)^{\frac{1}{2}}r/y$ be large is amply satisfied. But in fact the approximation (6-6) seems to be surprisingly good everywhere.

For any *constant* y, Eq. 6-6 is the approximate equation of a Mach line,[2] which shows that to this approximation the Mach lines are *parabolas*, with axes in the Mach direction of the undisturbed stream. (Oddly enough this is the same shape as was found, asymptotically, for the *shocks* in two-dimensional airfoil theory, in which, however, the Mach lines are straight.)

Of essential importance in the theory is the coefficient $kF(y)$ of $r^{\frac{1}{2}}$ in Eq. 6-6, which determines the curvature of the parabolic Mach lines. This is split up in Eq. 6-7 into a factor k depending only on the Mach number, in a way illustrated in Fig. E,6b, and a function $F(y)$ which is derived from the fundamental function $f(t)$ in a manner independent of Mach number. Thus for smooth slender bodies it depends on body shape alone. Mathematically F is the $(\frac{1}{2})$th derivative of f, multiplied by $\sqrt{\pi}$.

Now whereas for smooth slender bodies $F(y)$ is

$$\frac{1}{2\pi}\int_0^y \frac{S''(t)dt}{(y - t)^{\frac{1}{2}}} \tag{6-8}$$

an additional term arises when the surface has a discontinuity of slope, say at $x = t_1$, where the projectile radius is R_1, and $S'(X)$ changes by $(\Delta S')$. By a method similar to that used by Lighthill [*35*] to find the additional terms in the surface pressures (Art. 7), one can show that the additional term in $F(y)$ is

$$\left(\frac{\Delta S'}{2\pi}\right)\left[\frac{2}{(M_\infty^2 - 1)^{\frac{1}{2}}R_1}\right]^{\frac{1}{2}} h\left(\frac{y - t_1}{(M_\infty^2 - 1)^{\frac{1}{2}}R_1}\right) \tag{6-9}$$

[2] Only, of course, up to where it encounters a shock.

where the function $h(z)$ is plotted as a plain line in Fig. E,6c. It is zero for z less than -1, where it rises discontinuously to unity, and then falls off steadily, following closely the curve $(2z)^{-\frac{1}{2}}$ (dashed line) for $z > 3$. When this is so, the sum of Eq. 6-8 and 6-9 becomes effectively

$$\frac{1}{2\pi}\int_0^y \frac{dS'(t)}{(y-t)^{\frac{1}{2}}} \tag{6-10}$$

which is to be understood as including a term $(\Delta S'/2\pi)(y - t_1)^{-\frac{1}{2}}$ owing to the jump in $S'(t)$ at $t = t_1$. (This latter term could be written in the

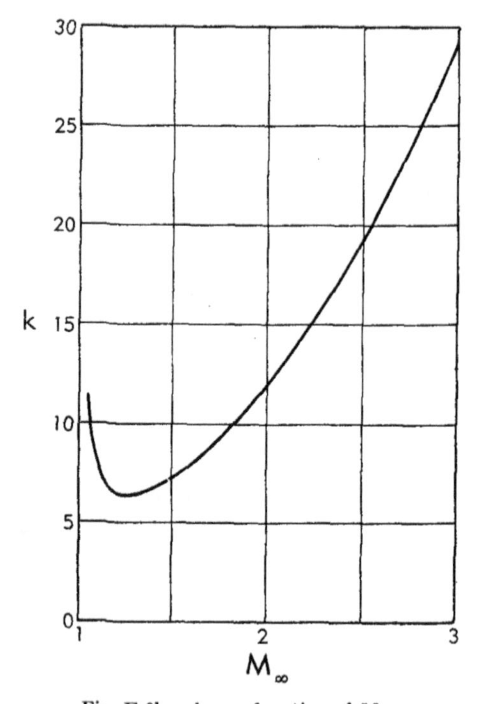

Fig. E,6b. k as a function of M_∞.

form of Eq. 6-9 with $h(z)$ given by the dashed line in Fig. E,6c. It is worth remarking that the total area between the two curves is zero.)

The discontinuity in $F(y)$, by an amount equal to the product of the first two factors in Eq. 6-9, takes place on the Mach lines given by $y = t_1 - (M_\infty^2 - 1)^{\frac{1}{2}}R_1$ through the point (t_1, R_1) where the slope of the surface changes. There are several such Mach lines although y has only one value, because (in Eq. 6-6 for the Mach lines) $F(y)$ may have any value between its two extremes. The Mach lines are a system of parabolas which concur to form a fan, just as they do in the Prandtl-Meyer expansion around a corner. Indeed the range of slopes dx/dr of these Mach lines, by Eq. 6-6, is

$$-\frac{1}{2}k(\Delta F)R_1^{-\frac{1}{2}} = -\frac{1}{2}k\left(\frac{\Delta S'}{2\pi}\right)2^{\frac{1}{2}}(M_\infty^2 - 1)^{-\frac{1}{4}}R_1^{-1} \tag{6-11}$$

and this is $\frac{1}{2}(\gamma + 1)M_\infty^4(M_\infty^2 - 1)^{-1}$ times the change in flow direction which gives welcome support to the theory, since it agrees with the first approximation to the two-dimensional flow around a corner.[3] The fan of Mach lines has physical reality only if $\Delta S' < 0$; otherwise of course there

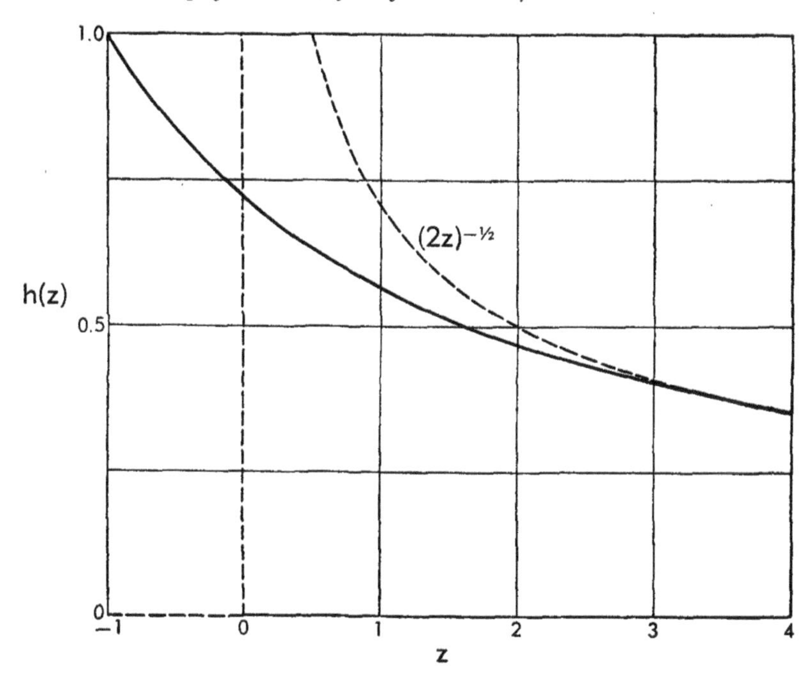

Fig. E,6c. The function $h(z)$.

is an attached shock; but its form can still be predicted from the form of $F(y)$, as will be seen below.

When there are several discontinuities of slope, say $t_1, t_2, \ldots,$ the expression (6-9) corresponding to each must be added to Eq. 6-8 to give $F(y)$. As indicated before, one can find $F(y)$ for a body whose slope or curvature changes too rapidly for Eq. 6-8 to be reliable, by simulating the region of rapid change by several (or a continuous distribution of) discontinuities of slope.

Using these considerations, the fundamental function $F(y)$ has been

[3] It is obvious that the immediate change at the surface must take place exactly as in two-dimensional flow, since the distances involved may be considered as infinitely small compared with the distance from the axis.

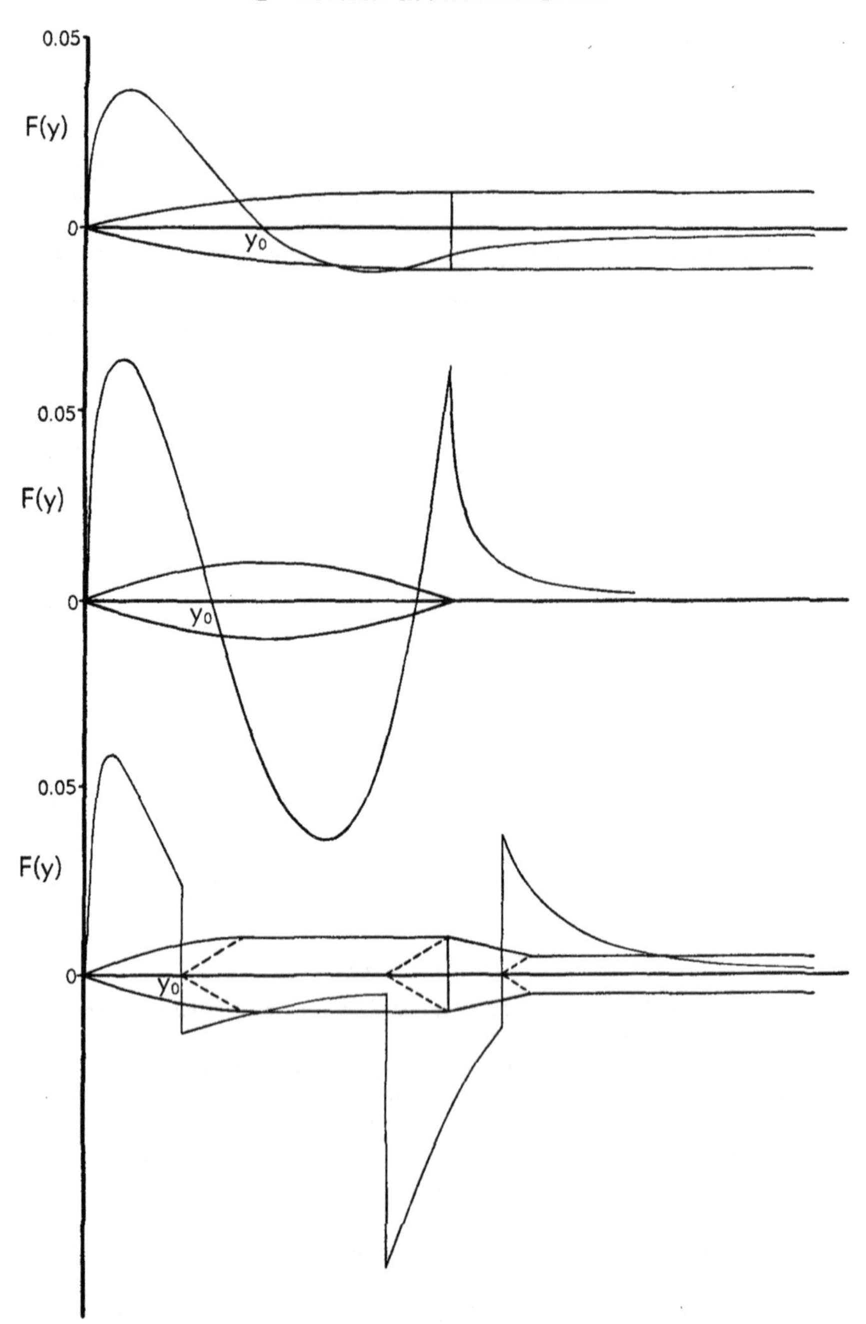

Fig. E,6d. The function $F(y)$ for three different body shapes.

calculated for three different body shapes, which are shown in Fig. E,6d with a graph of $F(y)$ superimposed on each. (In each case the length of the body is taken arbitrarily as unity.) The first is a shell shape ending in a cylindrical portion, with no allowance made for wake thinning. The second is the extreme case of a body pointed at both ends, which corresponds most closely to the airfoils of Art. 3 and 4; however, the linear theory (D,16; cf. Fig. D,16b) indicates for such bodies a very rapid compression ahead of the rear point, so that almost certainly the boundary layer separates, giving a true "mainstream" more as in Fig. E,6a. (On the other hand, if the boundary layer were controlled, the "trailing edge" might probably be a stagnation point, in a region of subsonic flow.) The third shape, which comes far closer to conditions met with in ordinary ballistics, incorporates three discontinuities in slope, one where the "5/10 caliber ogival head" is joined to the cylindrical portion, one where the wake leaves the cylindrical portion at the typical angle of 12°, and one where the wake straightens out. It would be slightly more realistic to make this straightening out rapid but continuous; then the jump in $F(y)$ would be similar. For this body both f and F depend on Mach number, even on linear theory; the calculations were done for $M_\infty = 2$. Fig. E,6d gives a general impression of the behavior of $F(y)$ for given maximum diameter; for variations of the maximum diameter, F is proportional to its square.

Notice that initially $F(y)$ is always positive. For that portion of the surface which is approximately conical (with semiangle δ, at the nose) $f(t)$ is approximately $\delta^2 t$, and so

$$F(y) \cong 2\delta^2 y^{\frac{1}{2}} \tag{6-12}$$

While $F(y)$ remains positive the corresponding parabolic Mach lines (through points on the axis a distance y from the nose) are, as Eq. 6-6 shows, convex to the oncoming stream. Where $F(y)$ is first zero, say for $y = y_0$, there is a straight Mach line, and those farther downstream are concave to the oncoming stream. In some cases there may be a further portion with $F(y) > 0$ again.

Now, when the condition that $(M_\infty^2 - 1)^{\frac{1}{2}} r/y$ is large (which was used to deduce Eq. 6-6) holds, i.e. on a given Mach line far downstream of its point of origin compared with the distance of the latter from the nose, a notable simplification to Eq. 6-3 for the disturbance velocities also results, giving

$$v = -(M_\infty^2 - 1)^{\frac{1}{2}} u = 2^{-\frac{1}{2}} (M_\infty^2 - 1)^{\frac{1}{4}} \frac{F(y)}{r^{\frac{1}{2}}} \tag{6-13}$$

Thus on any Mach line the disturbance velocity is ultimately perpendicular to the Mach line and proportional to the inverse square root of the distance from the axis. This is what would be expected of a conical sound wave, assuming that at large distances it leaves less and less tail behind it, and that its wave energy (proportional to the radius and the square of the amplitude) therefore becomes constant.

The information so far obtained, insofar as it relates to the potential flow field far from the axis, was deduced by two quite different methods in Whitham's first paper [*32*]. First he showed that the equation of continuity may be approximated with some appearance of reason, for large r, by the equation

$$2\left[\frac{\partial u}{\partial r} + (M_\infty^2 - 1)^{\frac{1}{2}}\frac{\partial u}{\partial x}\right] + \frac{u}{r} + \frac{(\gamma + 1)M_\infty^4}{(M_\infty^2 - 1)^{\frac{1}{2}}} u \frac{\partial u}{\partial x} = 0 \qquad (6\text{-}14)$$

of which the exact solution happens to be given by Eq. 6-6 and 6-13, with $F(y)$ an arbitrary function, which Whitham assumes to depend on the body shape. This solution at large distances from the axis may, for slender bodies, be compared with the linear theory (in some intermediate region where both are approximately valid) to determine $F(y)$ as in Eq. 6-7. Because he feels that the approximations leading to Eq. 6-14 are lacking in rigor, he then assumes a very general series expansion for u and v for large r, of the kind described at the end of Art. 1, and determines Eq. 6-6 and 6-13 as the first terms, respectively, in these series, by substituting the series in the exact equations of potential flow. Carried to one further approximation, the general behavior of supersonic axisymmetric potential flow at large distances from the axis is found to be given by

$$\begin{aligned} x - (M_\infty^2 - 1)^{\frac{1}{2}}r &= y - kF(y)r^{\frac{1}{2}} + k_1F^2(y)\ln r + \cdots \\ u &= -2^{-\frac{1}{2}}(M_\infty^2 - 1)^{-\frac{1}{4}}F(y)r^{-\frac{1}{2}} + k_2F^2(y)r^{-1} + \cdots \\ v &= -(M_\infty^2 - 1)^{\frac{1}{2}}u - \frac{(\gamma + 1)M_\infty^4}{4(M_\infty^2 - 1)}F^2(y)r^{-1} + \cdots \end{aligned} \qquad (6\text{-}15)$$

where k_1 and k_2 (like k) are specified functions of Mach number only, and the curves y = const are now *exact* Mach lines. Again the function $F(y)$ can be determined, for slender bodies, by comparison with linear theory. This systematic approach affords strong confirmation of the results obtained by the physical arguments given above. However, these latter arguments by themselves are sufficient to determine the shocks, as will be seen; it is therefore merely additional confirmation that in Whitham's earlier paper [*32*] the new terms, in $F^2(y)$, in Eq. 6-15 were not found to have any influence on the shock positions at large distances.

Before going on to find where the shocks cut off the potential flow field, it must be remarked that the conclusions obtained so far, and hence also what are to follow, are true not only for axisymmetric flows, but also, to a first approximation, for any flow about projectiles (not necessarily of revolution) which are slender in Ward's sense [*43*], i.e. make everywhere a small angle with the incident stream. For then (Art. 7) the flow, in regions where $(M_\infty^2 - 1)^{\frac{1}{2}}r/[x - (M_\infty^2 - 1)^{\frac{1}{2}}r]$ is not small, *is* approximately axisymmetric. For, if it is Fourier-analyzed with respect to the angular cylindrical polar coordinate θ, the coefficients of $\cos n\theta$ and $\sin n\theta$ are smaller than the axisymmetric term (independent of θ) by a factor of order the nth power of the fineness ratio, which is only canceled out near the body, where the fraction just mentioned *is* small and the coefficient of $\cos n\theta$ and $\sin n\theta$ behaves like r^{-n}. Further, the said axisymmetric flow is given by Eq. 6-3, with $f(t)$ related to the distribution of cross-sectional area downstream of the pointed nose exactly as for axisymmetric flow.

Thus, except near the projectile, the flow is roughly axisymmetric, and the Mach lines and shocks are where they would be for a symmetrically placed projectile of revolution with the same distribution of cross-sectional area. In particular, yaw makes little difference to the shocks on a slender projectile. They are yawed by only a small fraction of the projectile yaw. This is borne out by experiment and by more exact calculations on flow past yawing cones [*37*]. Little further reference to the case of nonaxisymmetric flow will be made in this article.

Determination of the front shock. The front shock will be determined first. This must lie ahead of the straight (approximate) Mach line through the nose, in the region where the left-hand side of Eq. 6-5 is negative. But y is essentially positive. Hence the integrals in Eq. 6-5 must be negative and of absolute magnitude exceeding y. This shows at once that the condition that $(M_\infty^2 - 1)^{\frac{1}{2}}r/y$ be large, on which Eq. 6-6 is (strictly speaking) based, must be amply satisfied. The latter equation is now used to determine the shock much as in the Friedrichs theory. Since (Art. 4) to a first approximation the direction of a shock bisects that of the Mach directions ahead of and behind it, the slope dx/dr of the shock must lie halfway between $(M_\infty^2 - 1)^{\frac{1}{2}}$ and the slope of the Mach line immediately behind it. Hence, by Eq. 6-6, along the shock,

$$\frac{d}{dr}\left[y - kF(y)r^{\frac{1}{2}}\right] = \frac{1}{2}\left[-\frac{1}{2}k\frac{F(y)}{r^{\frac{1}{2}}}\right] \tag{6-16}$$

The solution of this differential equation (which becomes linear when regarded as an equation for $r^{\frac{1}{2}}$ in terms of y), with the initial condition $r = 0$ when $y = 0$, is

$$r = \frac{4}{k^2}\frac{\left[\int_0^y F(t)dt\right]^2}{[F(y)]^4} \tag{6-17}$$

Thus the front shock is described parametrically by Eq. 6-6 and 6-17 (cf. Eq. 4-1 and 4-7 in the Friedrichs theory, which, it should however be remembered, goes to a second approximation).

Near the nose, or more precisely for that portion of the shock intersected by Mach lines springing from the portion of surface which is approximately conical, Eq. 6-12 holds, so that Eq. 6-6 and 6-17 become

$$x - (M_\infty^2 - 1)^{\frac{1}{2}} r = y - 2k\delta^2 (yr)^{\frac{1}{2}}, \qquad r = \frac{4}{9k^2\delta^4}\, y \tag{6-18}$$

whence

$$x = r[(M_\infty^2 - 1)^{\frac{1}{2}} - \tfrac{3}{4}k^2\delta^4] \tag{6-19}$$

Thus this forward portion of the shock is effectively straight and at an angle

$$(\tfrac{3}{4}k^2\delta^4)M_\infty^2 = \tfrac{3}{8}(\gamma + 1)^2 M_\infty^6 (M_\infty^2 - 1)^{-\frac{3}{2}}\delta^4 \tag{6-20}$$

to the undisturbed Mach direction. The strength of the shock in this region is therefore

$$\frac{\Delta p}{p} = \frac{3}{2}\gamma(\gamma + 1) M_\infty^6 (M_\infty^2 - 1)^{-1}\delta^4 \tag{6-21}$$

Notice that, by Eq. 6-18, this strength is maintained to a distance r from the axis such that the projectile remains practically conical over a distance $\frac{9}{4}k^2\delta^4 r$ from the nose; even if the latter distance be only small, the former distance may be quite large owing to the factor δ^4 multiplying r.

The correctness to a first approximation of Eq. 6-21 for the shock strength at the nose is checked by the analysis on the *exact* adiabatic theory of the axisymmetric flow past a cone of semiangle δ, which Taylor and Maccoll [*38*] first showed to be governed by ordinary differential equations only (H,16–19). The present author's proof [*36*] that Eq. 6-21 is, in this case, an approximation to the shock strength with an error at most of order δ^5 for small δ, is completely rigorous. (The true error, however, is more probably of about order δ^6.) Thus the theory is certainly correct for small enough δ. But how small need δ be for the approximation to be useful? This is answered by a comparison with numerical solutions of the ordinary differential equations mentioned above, which were computed by Kopal [*37*] at the Massachusetts Institute of Technology, and which, it should be mentioned, agree almost perfectly with experimental results. The comparison is made, on the angle between the shock and the undisturbed Mach direction, given approximately in Eq. 6-20, in Fig. E,6e. It is seen that, although Eq. 6-20 predicts a rather spectacularly weak shock compared with that produced by a symmetrically placed *wedge* of semiangle δ (for which the corresponding formula is $\frac{1}{4}(\gamma + 1)M_\infty^2(M_\infty^2 - 1)^{-1}\delta)$, nevertheless it overestimates the true shock strength as δ increases. The formula is satisfactory for $\delta < 10°$ (indeed for $\delta = 5°$ the curves would be indistinguishable) which is a rather larger range than for corresponding formulas in Art. 3 and 4. But the error is about 50 per cent for $\delta = 15°$. It must be stressed that, to a much greater extent for projectiles of revolution than for airfoils, *the range of practical shapes extends far outside the range in which any theories assuming small disturbances are useful;* in fact nose semiangles of 30° are common, and where this is so the flow can only be determined by numerical methods (Sec. H). On the other hand, future projectiles may be more slender, since the wave drag for given total volume varies (D,18) as the $\frac{8}{3}$ power of the fineness ratio. Notice that there is no

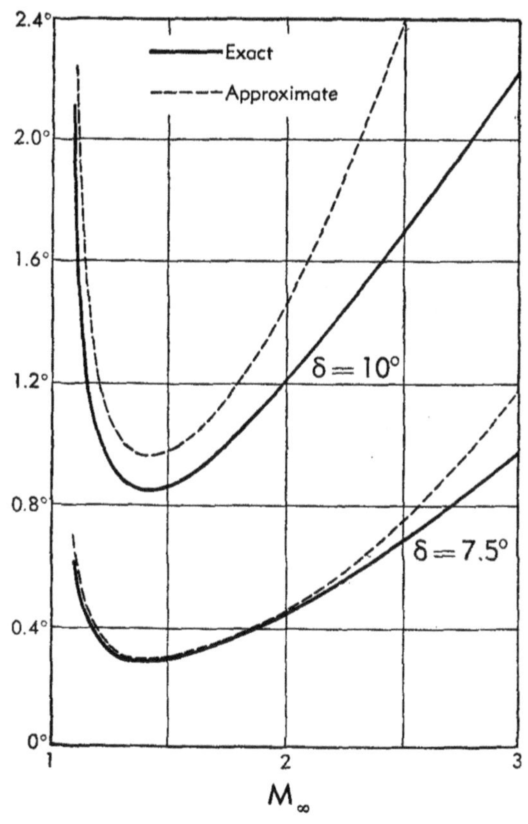

Fig. E,6e. Angle between shock and undisturbed Mach cone in axisymmetric flow at Mach number M_∞ past a solid cone of semiangle δ.

need to possess Eq. 6-20 and 6-21 for the cone, since accurate tables are available. Their essential purpose is to make possible the above comparison with exact adiabatic theory. Note also, as pointed out after Eq. 6-3, that the Whitham theory, even when inaccurate near the nose, is *still correct outside a cylinder of sufficient radius.*

Returning to the formula (6-17) for the distance from the axis at which the Mach line, emanating from the axis at a distance y from the nose, encounters the shock, notice that $F(y)$ becomes an increasingly larger multiple of $r^{-\frac{1}{2}}$, but that the coefficient cannot increase without limit because (Fig. E,6d) the maximum of the integral is

$$\int_0^{y_0} F(t)dt.$$

For very large r, $F(y)$ therefore becomes very small, and so $y \to y_0$. Thus all Mach lines ahead of the straight Mach line $y = y_0$ intersect the front shock, but the straight Mach line continues all the way to infinity. For very large r, with $y \cong y_0$, the integral in Eq. 6-17 has reached its maximum and combining Eq. 6-17 with Eq. 6-6,

$$x - (M_\infty^2 - 1)^{\frac{1}{2}} r \cong y_0 - \left[2k \int_0^{y_0} F(t)dt\right]^{\frac{1}{2}} r^{\frac{1}{4}} \tag{6-22}$$

Thus the front shock stands ahead of the straight Mach line $y = y_0$ by an amount asymptotically proportional to the *fourth* root of the distance from the axis (and, as indicated in the first sentence of this paragraph, by rather less than Eq. 6-22 gives for smaller r); this increases much less rapidly than the corresponding quantity for two-dimensional airfoil flow, which is proportional to the inverse *square* root of the distance (Art. 4).

The shock strength is deduced, either from the angle between the shock and the undisturbed Mach lines (obtained by differentiating Eq. 6-22), or from the expressions (6-13) for the disturbance velocities behind it, as

$$\frac{\Delta p}{p} \sim \gamma \left(\frac{2^{\frac{1}{2}}}{\gamma + 1}\right)^{\frac{1}{2}} (M_\infty^2 - 1)^{\frac{1}{8}} \left[\int_0^{y_0} F(t)dt\right]^{\frac{1}{2}} r^{-\frac{3}{4}} \tag{6-23}$$

for large r. For similar bodies, of different lengths l and fineness ratios ϵ, this expression for the shock strength at "miss-distance" r (from the path of the projectile) is proportional to $(r/l)^{-\frac{3}{4}}$ and to ϵ, while depending little on Mach number. It would be reasonable, if the shock strength at a large distance from a projectile not of revolution, and not even slender, had to be estimated, to hold to this result but with fineness ratio interpreted as (frontal area)$^{\frac{1}{2}}$/length. Using the curves in Fig. E,6c to estimate the integral in Eq. 6-23 in terms of frontal area and length, the rough order-of-magnitude formula for the shock strength far from the line of flight,

$$\frac{\Delta p}{p} \cong \epsilon \left(\frac{l}{r}\right)^{\frac{3}{4}} \tag{6-24}$$

would in such cases be used.

The asymptotic proportionality of the strength of ballistic shocks to the inverse $\frac{3}{4}$ power of the miss-distance was first observed experimentally, and also explained theoretically, by DuMond, Cohen, Panofsky, and Deeds [*39*]. For the bullets and shells used by them the $\frac{3}{4}$ power law was closely verified for miss-distances exceeding about 1,000 projectile diameters, while the rate of fall of strength with distance was rather more rapid closer to the projectile (as the above theory indicates). Two theoretical arguments explaining the law were given by these authors, of which the more convincing is an energy method which relates the rate of loss of wave energy between the shocks to the rate of gain of entropy of the fluid. It is very satisfactory that this gives the same conclusion as Whitham obtained by his quite different approach (which does not need to consider entropy variation); of course the reader should no longer regard such a fact as paradoxical, after the full discussion of the analogous two-dimensional problem in Art. 4.

Determination of the rear shock. We pass to the determination of the rear shock. This will not normally be attached, since the body-wake combination is not expected to have a discontinuity in slope concave to the stream. Rather the rear shock will originate in the midst of the flow, as a result of the running together of Mach lines, in the manner described in Art. 3. Photographs show much branching at the base of such shocks, as in the delta of a great river. This is incidental, being due to irregularities propagated along the Mach lines. The present theory applies to the form of the shock only where it becomes a recognizable single branch.

The flow is nonuniform both ahead of and behind the rear shock. If, at any point of it, the downstream-pointing Mach lines ahead of and behind it are those with the values y_1 and y_2 (respectively) for y, then, since the slope of the shock (to a first approximation) bisects that of these two Mach lines, Eq. 6-6 gives

$$\frac{d}{dr}[y - kF(y)r^{\frac{1}{2}}] = \frac{1}{2}\left[-\frac{1}{2}kF(y_1)r^{-\frac{1}{2}} - \frac{1}{2}kF(y_2)r^{-\frac{1}{2}}\right] \qquad (6\text{-}25)$$

along the shock. Here the subscript on y on the left-hand side is immaterial because

$$y_1 - kF(y_1)r^{\frac{1}{2}} = y_2 - kF(y_2)r^{\frac{1}{2}} \qquad (6\text{-}26)$$

since the two Mach lines meet on the shock.

Writing Eq. 6-26 as

$$r^{\frac{1}{2}} = \frac{y_2 - y_1}{k[F(y_2) - F(y_1)]} \qquad (6\text{-}27)$$

and differentiating this, both r and dr may be eliminated from Eq. 6-25, giving an equation which may be put in the form

$$F(y_2)dy_2 - F(y_1)dy_1 = \tfrac{1}{2}d\{(y_2 - y_1)[F(y_1) + F(y_2)]\} \qquad (6\text{-}28)$$

This form was ingeniously selected by Whitham to permit direct integration as

$$\int_{y_1}^{y_2} F(t)dt = \tfrac{1}{2}(y_2 - y_1)[F(y_1) + F(y_2)] \qquad (6\text{-}29)$$

where no additive constant appears since no infinity in F can be supposed where $y_1 = y_2$ (i.e. where the shock originates).

Eq. 6-29 has a simple geometric interpretation which makes it easy, in practice, to evaluate the relation between Mach lines y_1 and y_2 which intersect on the rear shock. The left-hand side is the area under the curve $F(t)$ for $y_1 < t < y_2$. The right-hand side is the area under the straight segment joining the points y_1, $F(y_1)$ and y_2, $F(y_2)$. Hence for Eq. 6-29 to be satisfied, there must be parts of the graph of $F(y)$ lying both above and below the straight segment, and the two lobes which lie between the straight segment and the curve must have equal area. This is illustrated in Fig. E,6f, where the lobes are shaded. Further, by Eq. 6-27, the *slope* of the said straight segment is equal to $k^{-1}r^{-\frac{1}{2}}$, i.e. it is proportional to

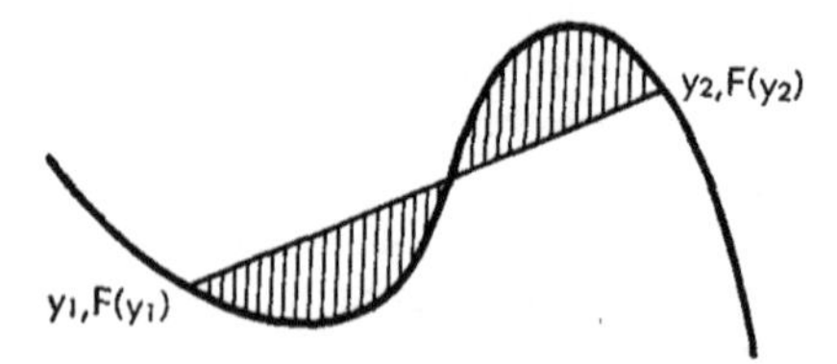

Fig. E,6f. The relation between Whitham's y_1 and y_2.

the inverse square root of the distance from the axis of the point on the shock where the two Mach lines meet. This shows that this slope must be positive; also, it gives the value of r at the corresponding point on the shock. From Eq. 6-6, the x coordinate follows, and, by varying the straight segment so that the lobes remain of equal area, the whole rear shock can be plotted.

A rear shock will *originate* on a Mach line specified by a value $y = y_i$ for which the slope of the graph of $F(y)$ is positive and locally a maximum, so that r is least (i.e. the point on the shock nearest the axis is being considered). When $F(y)$ is a smooth curve, y_i must be a point of inflection. This is illustrated in Fig. E,6g, where successive straight segments joining y_1, $F(y_1)$ to y_2, $F(y_2)$ and cutting off lobes of equal area are also shown, the function $F(y)$ being chosen as the simplest of those graphed in Fig. E,6d. As the slope decreases (so that we go farther out along the rear shock) it is clear that $y_1 \to y_0$ and $y_2 \to \infty$. Thus all the Mach lines behind the straight Mach line $y = y_0$ meet the rear shock: Those with $y_0 < y < y_i$ meet it in front, and those with $y_i < y$ meet it behind. The general properties of the shock system will be inferred from Fig. E,6g. Any slight alterations resulting from functions $F(y)$ of more complicated shape will be discussed later.

One may suppose that $F(y) \to 0$ as $y \to \infty$, if one takes the effective wake to be ultimately cylindrical. Then the behavior of the rear shock at very large distances from the axis, corresponding to y_1 near y_0 and large y_2, is given by approximating Eq. 6-27 and Eq. 6-29 as

$$r^{\frac{1}{2}} = \frac{y_2 - y_1}{-kF(y_1)}, \qquad \int_{y_0}^{\infty} F(t)dt = \frac{1}{2}(y_2 - y_1)F(y_1) \qquad (6\text{-}30)$$

Substituting for $y_2 - y_1$ in the first equation from the second, and taking the positive square root (noticing from Fig. E,6d that $F(y_1) < 0$) we get

$$-F(y_1) = \left[\frac{-2\int_{y_0}^{\infty} F(t)dt}{kr^{\frac{1}{4}}}\right]^{\frac{1}{2}} \tag{6-31}$$

whence by Eq. 6-6 the asymptotic form of the rear shock is

$$x - (M_\infty^2 - 1)^{\frac{1}{2}}r \cong y_0 + [-2k\int_{y_0}^{\infty} F(t)dt]^{\frac{1}{2}}r^{\frac{1}{4}} \tag{6-32}$$

This asymptotic form should be compared with that of the front shock Eq. 6-22. It is seen at once that the rear shock stands behind the

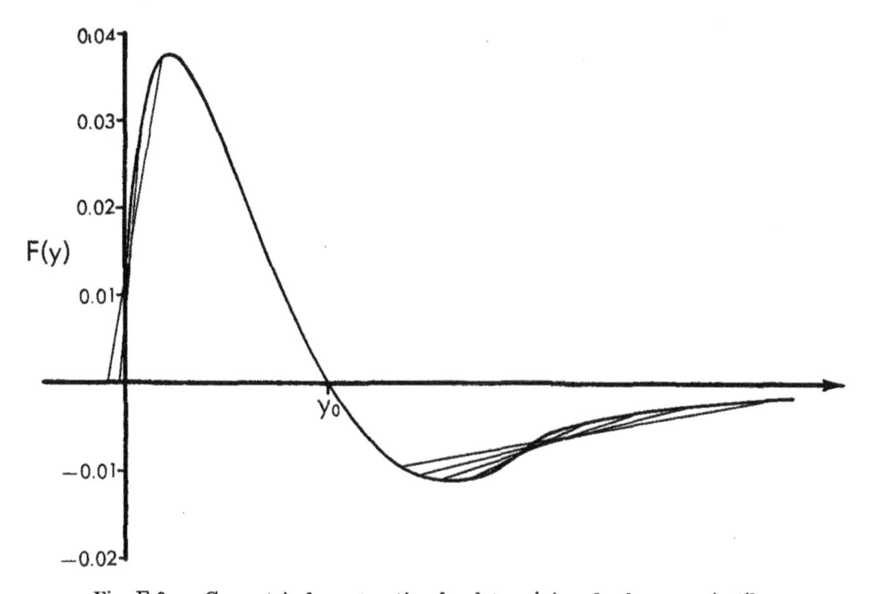

Fig. E,6g. Geometrical construction for determining shocks on projectiles.

straight Mach line $y = y_0$ by an amount proportional to $r^{\frac{1}{4}}$, just as the front shock so stands ahead of it. But, in fact, *the constants of proportionality are the same* in Eq. 6-22 and 6-32. This is because it may be shown that, because $F(y)$ as defined in Eq. 6-7 is a sort of derivative, the integral

$$\int_0^{\infty} F(t)dt$$

is zero (assuming that $f(0) = 0$ and $f = O(y^{-1})$ as $y \to \infty$). Thus at very large distances the front and rear shocks are of equal strength. Now the experiments [*39*] support this conclusion even at miss-distances of about 100 projectile diameters, for which the $-\frac{3}{4}$ power law has not yet come into operation. That this should be so is explained by evaluating the shock paths to the next approximation at large miss-distances, on which they are

$$x - (M_\infty^2 - 1)^{\frac{1}{2}} r \cong y_0 \mp \left[2k \int_0^{y_0} F(t)dt\right]^{\frac{1}{2}} \left[r^{\frac{1}{4}} + \frac{r^{-\frac{1}{4}}}{2kF'(y_0)}\right] \quad (6\text{-}33)$$

showing that, as observed, the front and rear shock strengths should equalize out before they begin to satisfy the $-\frac{3}{4}$ power law accurately. The *length* of the N wave is, at large distances from the axis, twice the second term in Eq. 6-33. The experiments confirmed this by showing that the slope of length against distance on logarithmic scales is close to 0.25, but always less if anything.

It may be remarked that the arguments used to determine the rear shock apply equally to the front shock, for which, however, the Mach lines ahead are straight, so that effectively $F(y) = 0$. It is still true that straight segments cutting off lobes of equal area (see the left portion of Fig. E,6g) join values of y corresponding to Mach lines meeting at a point on the shock, whose distance from the axis is proportional to the inverse square of the slope of the segment. It is because $F = 0$ for the left-hand value of y that this geometrical relation between r and y has for the front shock a relatively simple analytical form (6-17). But in practice the geometrical relation may be easier for purposes of calculation.

We next consider the pressure variations measured by a suitable instrument as a supersonic projectile passes, missing it by a distance r large compared with the length of the projectile. By Eq. 6-13 the pressure is given by

$$p - p_\infty = \rho_\infty q_\infty^2 2^{-\frac{1}{2}} (M_\infty^2 - 1)^{-\frac{1}{4}} F(y) r^{-\frac{1}{2}} \quad (6\text{-}34)$$

and it is the variation of this with x for constant r in which we are interested, where x and y are related by Eq. 6-6. It is seen at once that the pressure gradient $\partial p/\partial x$ (which when multiplied by q_∞ gives the time rate of change of the observed pressure) is

$$\frac{\partial p}{\partial x} = \rho_\infty q_\infty^2 2^{-\frac{1}{2}} (M_\infty^2 - 1)^{-\frac{1}{4}} \frac{F'(y) r^{-\frac{1}{2}}}{1 - kF'(y) r^{\frac{1}{2}}} \quad (6\text{-}35)$$

The denominator in Eq. 6-35 marks a difference between the linear and Whitham theories by expressing the change in pressure gradient due to the spreading out or bunching in of the Mach lines. It never becomes negative (this would involve encountering a "limit line") as shocks always appear before this can happen.

Now, between the shocks at large distances from the axis, y is close to y_0, and so the pressure gradient (6-35) is practically constant, giving an "N wave" form of pressure curve (Fig. E,4a). Further, when r is much larger than $[kF'(y_0)]^{-2}$, only the second term in the denominator matters. When this is so, Eq. 6-35 becomes (using the expression (6-7) for k)

$$\frac{\partial p}{\partial x} = -\rho_\infty q_\infty^2 \frac{(M_\infty^2 - 1)^{\frac{1}{4}}}{(\gamma + 1) M_\infty^4} \frac{1}{r} \quad (6\text{-}36)$$

This ultimate slope of the N is independent of body shape and inversely proportional to miss-distance. The result is very similar to that obtaining in two-dimensional airfoil theory, for which $\partial p/\partial x$ is exactly twice as much (as follows from Eq. 4-13); actually it is the one point for which there is such quantitative similarity. Eq. 6-36 means that at 15°C an observer will record a time rate of pressure fall (after the initial jump at the shock) equal to about

$$0.2\,\frac{(M_\infty^2 - 1)^{\frac{1}{2}}}{M_\infty}\quad\frac{\text{atmospheres per millisecond}}{\text{miss-distance in meters}} \tag{6-37}$$

In Eq. 6-37 the second factor is not far from 1 if the projectile speed is well supersonic. Hence an instrument that could measure this time rate of pressure fall between the shocks would give at once the miss-distance of a supersonic projectile to a good approximation even if its Mach number were not known. (Notice that Eq. 6-37 is true independently of any approximation for slender bodies, provided that the miss-distance is large enough, since the function $F(y)$ does not appear therein. Notice also that, since Eq. 6-36 is asymptotically true both for projectiles of revolution and, within a factor of 2, for the very extreme case of two-dimensional flow, it may be expected to be true to good approximation for projectiles of any shape.)

We pass to the consideration of the pressure behind the rear shock. Here $F'(y)$ is always considerably smaller than $k^{-1}r^{-\frac{1}{2}}$, since the geometry of Fig. E,6f shows that the straight segments (of slope $k^{-1}r^{-\frac{1}{2}}$) always intersect the curve from below. Hence the argument leading to Eq. 6-36 is completely falsified. In fact the pressure tails off rapidly behind the rear shock. By considering Eq. 6-7 for large y one sees that

$$F(y) \sim -f_0 y^{-\frac{3}{2}}, \quad \text{where } f_0 = \tfrac{1}{2}\int_0^\infty f(t)dt \tag{6-38}$$

For slender bodies f_0 is the effective cross-sectional area of the wake, divided by 4π. Using the approximation (6-38) one may calculate the pressure immediately behind the rear shock, from Eq. 6-34 and 6-30, as

$$p - p_\infty = -\rho_\infty q_\infty^2 2^{-\frac{1}{2}}(M^2 - 1)^{-\frac{1}{4}} f_0 \left[\frac{2}{k}\int_0^{y_0} F(t)dt\right]^{-\frac{3}{4}} r^{-\frac{7}{8}} \tag{6-39}$$

for large r. This is essentially negative, and of a slightly smaller order of magnitude than the value ($\sim r^{-\frac{3}{4}}$) immediately ahead of the shock. Also, since, for constant r, $dx/dy = 1 - kF'(y)r^{\frac{1}{2}}$ does not, as remarked above, depart greatly from 1, the subsequent behavior of the pressure as x increases is similar to its behavior as y increases, namely, a subsidence to zero from below, in proportionality (6-38) to $y^{-\frac{3}{2}}$.

To illustrate the great difference between the flow pattern near the body and that at large distances, the pressure is graphed against x for two fixed values of r in Fig. E,6h. The lower curve corresponds to a value of r soon after the appearance of the rear shock, which is still much weaker than the front shock. There is still a region of compression behind the front shock, and a sizeable tail behind the rear shock (indeed the predictions of linear theory are still very roughly correct in this region). The upper curve corresponds to a very large value of r, for which the asymptotic formulas are valid. The N wave dominates this pressure curve, and the tail is reduced in relation to it.

Additional shocks. We must now consider what modifications to the above theory are necessary when, as usually in practice (Fig. E,6d), the graph of $F(y)$ is more complicated than that used for illustration in

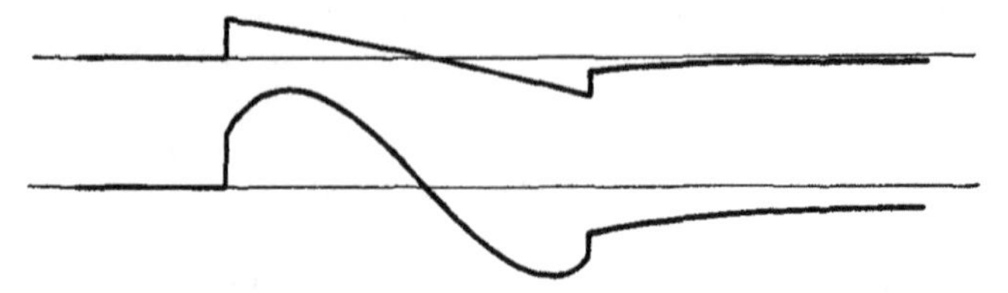

Fig. E,6h. Pressure signatures, far from and near a projectile, with time scales adjusted to make shocks coincide.

Fig. E,6g. First consider the effect of a "kink" in the body surface, producing a kink in the graph of $F(y)$ (Fig. E,6i) so that it has two points of inflection with positive slope. Then a shock will start from points corresponding to each of the two points of inflection. A typical point on the main rear shock will be represented by the straight segment BC. At first sight it would appear that the Mach line corresponding to B also meets the kink-produced shock, at a point represented by the straight segment AB. But the value of r corresponding to AB exceeds that corresponding to BC, and so there is no real encounter of the Mach line with the kink-produced shock before the former is cut off by the main rear shock. On the other hand the Mach line corresponding to E is cut off by the kink-produced shock, at a point represented by DE, before it can meet the rear shock at a point represented by EF. Between E and B there must be a point Y which is on a complete straight segment XYZ, such that both XY and YZ cut off lobes of equal area. Then both XY and YZ correspond to the same point, which is clearly where the two shocks run together. Above this they unite to form a single shock, whose progress may be followed by the use of single straight segments X_1Z_1 (with X_1 to the left of X, Z_1 to the right of Z) such that the total area of curve on both sides is the same. (However, it should be remarked that one can also draw kinks which produce no shock, because the rear shock cuts off *all* the associated Mach lines before they can run together; this means geometrically that no line such as DEF exists.)

Next consider a discontinuous function $F(y)$, such as the third curve in Fig. E,6d, corresponding to a shell with (somewhat idealized) wake thinning. The straight Mach line $y = y_0$, $F(y) = 0$, is seen to originate in the middle of the fan emanating from the shoulder. Behind this there are three places where the slope of the curve has locally a positive maximum. Of these the most important, connected with the origin of the main rear shock, is the third discontinuity, where the slope of the curve is $+\infty$. Obviously a series of straight segments can be drawn, crossing this vertical line and cutting off lobes of the curve with equal area on both sides, with slopes from $+\infty$ downward. They correspond to a shock,

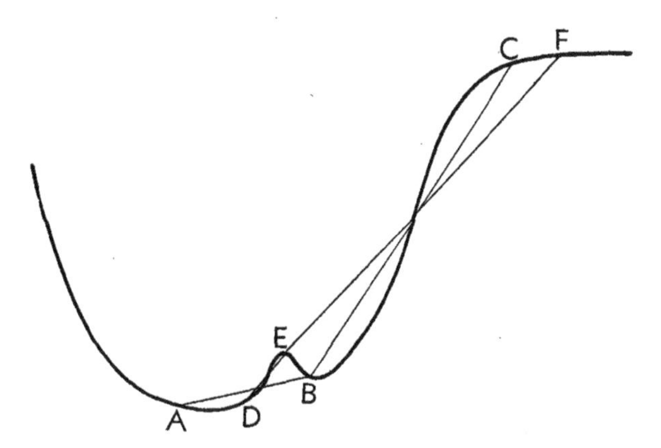

Fig. E,6i. Construction for determining bifurcated rear shock.

attached to the wake at the point where it straightens out. (If a wake shape which straightened out more gradually had been used, then $F(y)$ would have a steep but not vertical maximum of slope in this neighborhood and the shock would then be detached.) There is little reason to doubt the accuracy of this prediction of an attached shock in a corner, since the imaginary fan of Mach lines which it replaced was predicted correctly (see above).

But immediately behind each of the other discontinuities in $F(y)$ (where the slope of the curve is $-\infty$) there is a local positive maximum of slope which may correspond to a shock forming on the rear Mach line of the Prandtl-Meyer fan associated with the expansion around shoulder and rim respectively. In the case of the shoulder such a shock obviously forms, though with a value about 0.3 for the slope $k^{-1}r^{-\frac{1}{2}}$, which for $M_\infty = 2$ ($k = 12$) corresponds to a distance r from the axis of 7.7 projectile lengths. It gradually runs into the rear shock, exactly as discussed above for kink-produced shocks. But actually no shock forms on the rear Mach line emanating from the rim, because it is cut off by the rear shock first; careful measurement shows in fact that no such line as *DEF* (Fig. E,6i) can exist here.

The front part of the complete flow pattern in a meridian plane, as calculated by Whitham using his geometrical method, is illustrated in Fig. E,6j. The shocks are shown as heavy lines, and certain Mach lines are also shown, namely, in front and rear of each fan and also the straight Mach line in the middle of the forward fan.[4] The general pattern is of the form familiar from bullet photographs. Actually the strength of the front shock must be overestimated (see Fig. E,6d; actually the nose semi-angle is about 20°), but if a reliable estimate were required for practical

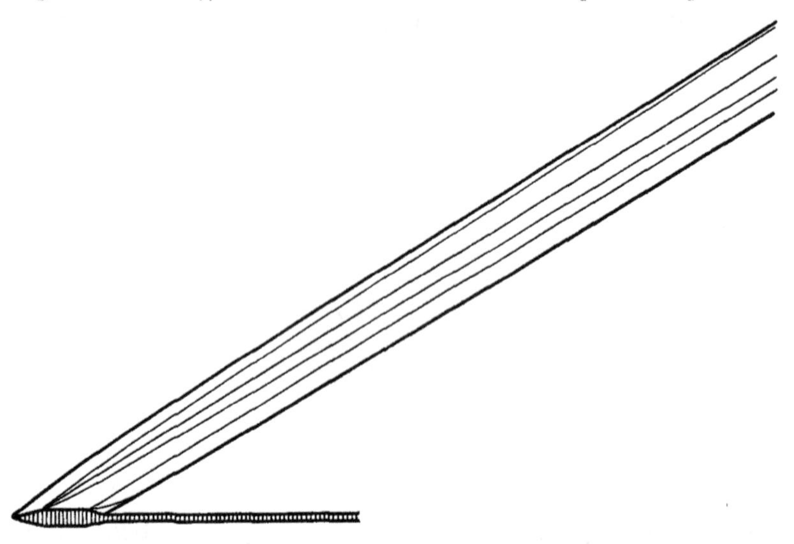

Fig. E,6j. Theoretically determined pattern of flow past a shell.

purposes, this error could be allowed for by adjusting $F(y)$ near the nose so that the deduced nose shock strength was as given by the numerical calculations for the cone. Similarly, accuracy could be improved near the base of the rear shock, making it detached by giving $F(y)$ a rapid but not discontinuous rise where the wake straightens out.

Drag related to downstream flow. Finally, the relation between the wave drag of the projectile and the flow behind it will be considered briefly, eking out, from analogies with the two-dimensional theory (Art. 4), our less precise knowledge of the flow. First we relate the drag to the flux of wave energy (or, what is essentially the same, to the flux of additional longitudinal momentum) across a plane far downstream and perpendicular to the axis. As in Art. 4, the wave energy of the N wave (being proportional to the wave length times the radius, i.e. to $r^{5/4}$, multiplied by the square of the amplitude, i.e. by $r^{-3/2}$) tends to zero far downstream (like $r^{-1/4}$), and the same is true, but more so (like $r^{-1/2}$), of the tail.

[4] The shock which forms on the rear Mach line of the forward fan is just off the picture; this hardly matters since it is not a phenomenon of any consequence.

Only the term Ts in the wave energy contributes, in the limit, to the total flux, and we have (cf. Eq. 4-43)

$$\text{Drag} = \int_0^\infty \rho_\infty Ts\, 2\pi r dr \tag{6-40}$$

where only the square of the specific entropy is neglected.

Since specific entropy is conserved for any fluid element, except where it increases at the shocks by an amount related to the deflection by Eq. 3-20, the integral in Eq. 6-40 may be evaluated to a first approximation by taking it over both shocks, with s signifying the *gain* in specific entropy across each. Thus

$$\begin{aligned}\text{Drag} &= \int_{\text{both shocks}} \rho_\infty T c_p \frac{\gamma^2 - 1}{12} \frac{M_\infty^6}{(M_\infty^2 - 1)^{\frac{3}{2}}} (\Delta v)^3 2\pi r dr \\ &= \rho_\infty q_\infty^2 \frac{\gamma + 1}{12} \frac{M_\infty^4}{(M_\infty^2 - 1)^{\frac{3}{2}}} 2\pi \left\{ \int_{\text{front shock}} 2^{-\frac{3}{2}} (M_\infty^2 - 1)^{\frac{3}{4}} \frac{F^3(y)}{r^{\frac{3}{2}}} r dr \right. \\ &\qquad \left. + \int_{\text{rear shock}} 2^{-\frac{3}{2}} (M_\infty^2 - 1)^{\frac{3}{4}} \frac{[F(y_2) - F(y_1)]^3}{r^{\frac{3}{2}}} r dr \right\}\end{aligned} \tag{6-41}$$

where the nondimensional velocity v has been substituted from Eq. 6-13. By using Eq. 6-17, and the expression (6-7) for k, the contribution from the front shock is easily evaluated as

$$\rho_\infty q_\infty^2 \pi \int_0^{y_0} F^2(y) dy \tag{6-42}$$

By using Eq. 6-27, and also (in a sufficiently cunning manner) Eq. 6-28, the contribution from the rear shock can actually be put into the same form, but with the limits y_0 and ∞. Hence, combining these,

$$\text{Drag} = \rho_\infty q_\infty^2 \pi \int_0^\infty F^2(y) dy \tag{6-43}$$

If $F(y)$ is substituted from Eq. 6-7, Eq. 6-43 reduces to Kármán's familiar form (D, 18; Eq. 18-9)

$$\rho_\infty q_\infty^2 \pi \int_0^\infty \int_0^\infty f'(x) f'(y) \ln \frac{1}{|x - y|} dx dy \tag{6-44}$$

This is yet another check on the correctness of Whitham's theory.

Other satisfactory points are (1) the simple form of the drag when expressed in terms of the function $F(y)$ which is fundamental in the theory, and (2) the fact that the part of the integral (6-43) shown in Eq. 6-42 corresponds to drag associated with degradation of energy at the front shock, while the rest is similarly associated with the rear shock.

When the function $F(y)$ has been defined in terms of a distribution of cross-sectional area which decreases in the wake to an ultimate value less than the base area of the projectile, the drag (6-43) includes more than what is *normally* called the wave drag, namely, the resultant of the normal forces on the head associated with the difference of the pressure from its undisturbed value. For it includes, in addition, the suction forces on the thinning wake. The resultant of these forces is clearly that part of the base drag which acts outside the cylinder enclosing the ultimate wake. The remainder of the base drag, which may reasonably be called "form drag," and also the skin friction drag, are associated with flux of reduced longitudinal momentum in the turbulent wake itself, due essentially to *shear* effects. It would seem reasonable to use "wave drag" (except where there is danger of confusion with established usage) to signify the drag due to the appearance of the shocks (as a result of *compression* effects), a drag whose influence on the fluid behind the projectile is spread out very widely indeed and correspondingly thinly.

Of course the pressure drag on the head *can* be expressed in the form (6-42), provided only that $F(y)$ signifies its value for the same projectile with *unthinned* cylindrical wake. It is an interesting consequence that the difference between the expressions (6-42) for thinned and unthinned wake is a known proportion of the base drag. Since the base drag can be calculated independently from the angle through which the main stream turns at the rim (which must already have been used in calculating $F(y)$ for the shell with thinned wake), it appears that a check is here possible which may in further developments of the theory permit still more accurate values of $F(y)$, and so of the flow pattern to be inferred.

If the drag is compared with the wave energy flux across a plane not so far downstream, the contribution from entropy is smaller because the whole of either shock does not lie ahead of the plane. The deficiency is made up from the wave energy in the N wave and tail. Since, as remarked above, this is asymptotically proportional to $r^{-\frac{1}{4}}$, it falls off at a rate proportional to $r^{-\frac{5}{4}}$. This is balanced by the rate of entropy gain (proportional to the cube of the shock strength, and so to $r^{-\frac{9}{4}}$, but multiplied by $2\pi r$ to allow for axisymmetry). The details are omitted.

Some final remarks are needed on the mass flow behind the projectile, since (Art. 4) positive work is done by the pressure of the atmosphere in increasing this anywhere, and negative work in decreasing it. It is easiest to calculate the first approximation to the mass flow in the N wave and tail across a *cylinder* of radius r, coaxial with the projectile. This, by the expression (6-13) for v, is proportional to the integral of $F(y)$ with respect to x for constant r, taken between the shocks and behind the rear shock. By Eq. 6-6 this is

$$\int F(y)[1 - kF'(y)r^{\frac{1}{2}}]dy \qquad (6\text{-}45)$$

Now Eq. 6-45 would be zero if the limits were 0 and ∞. Hence it may be replaced, with the sign changed, by the same integral taken over the intervals previously excluded from the range of integration, i.e. from zero to the value of y on the front shock, and from y_1 to y_2 in the notation used for the rear shock. That the integral over each of these intervals vanishes follows trivially from Eq. 6-17 for the front shock and Eq. 6-27 and 6-29 for the rear shock.

To the first order then, the mass flow is due to Whitham's approximate velocity field (6-13) and its total is zero. Note that to achieve this for large r it is essential that the N wave becomes balanced. Indeed the mass outflow through a cylinder of very large radius r, in the front half of the N wave, due to a disturbance velocity v proportional to $r^{-\frac{3}{4}}$ acting across an area proportional to $r \cdot r^{\frac{1}{4}}$, is *increasingly large* as the cylinder increases in radius. It is balanced however by the mass inflow in the rear half. (There is nothing extraordinary in this, which is equally true of any half wavelength in a harmonic train of spherical sound waves.) We have no second approximation to the flow, but we can be sure that, as in the two-dimensional case, there is, to a second approximation, a net mass outflow far from the axis to balance the reduced mass outflow behind the projectile due to increased entropy. One may also expect that pressure waves produced by reflection from the front shock and transmission through the rear shock play a part in this balance.

E,7. Supersonic Projectile Theory: Surface Pressures. In this article, as foreshadowed in Art. 6, we discuss fully the errors in the linear theory of the surface pressures on a supersonic projectile, and briefly describe the methods of Broderick [*34*] and Van Dyke [*54*] for proceeding to a second approximation for the pressure on projectiles of revolution in axisymmetric flow, and the author's analogous work [*40*] on the effect of yaw.

In the early work on the small perturbation theory of supersonic flow, it was gradually realized that, in its application to axisymmetric flow past projectiles, there were two peculiar features, as follows: (1) The axial and radial disturbance velocities u and v, given by

$$\begin{aligned} u &= -\int_0^{x-(M_\infty-1)^{\frac{1}{2}}r} \frac{f'(t)dt}{[(x-t)^2-(M_\infty^2-1)r^2]^{\frac{1}{2}}} \\ v &= \frac{1}{r}\int_0^{x-(M_\infty-1)^{\frac{1}{2}}r} \frac{(x-t)f'(t)dt}{[(x-t)^2-(M_\infty^2-1)r^2]^{\frac{1}{2}}} \end{aligned} \qquad (7\text{-}1)$$

(cf. Eq. 6-3), are *not of the same order of magnitude* near the projectile surface. In fact Eq. 7-1 shows that $-v/u$ is a sort of average of $(x-t)/r$ for $0 < t < x - (M_\infty^2 - 1)^{\frac{1}{2}}r$; this average would be expected to be large on the surface of a slender projectile. Indeed the square of v is normally comparable with u near the surface, and hence the question arises *whether the use of a linear theory is justifiable, since it predicts u by neglecting v^2*. (2) While on the one hand the calculation of the fundamental function

$f(t)$, by satisfying exactly the boundary condition at the surface, is a very lengthy process indeed, it is not at all obvious under what conditions the approximation $f(t) = S'(t)/2\pi$, which is obtained from the boundary condition at the surface by neglecting consistently the squares of the disturbances, is justified.

These two questions were tackled by the author [*42*] in 1945, by working out the order of magnitude of the next approximation to the flow (which Broderick and Van Dyke were later to calculate explicitly). First he showed that the errors in predictions of u and v near the surface by use of the linear equation of motion take the form of a factor differing from unity by not more than the order of magnitude of u itself. (Work equivalent to this investigation will be presented below.) But, as a consequence, the surface pressure distribution must be calculated from the equation

$$C_p = -2u - v^2 \tag{7-2}$$

not the equation $C_p = -2u$ which is usual in the linear theory, because as stated above v^2 is comparable with u in magnitude near the surface.

Secondly, the author showed that the approximation $f(t) = S'(t)/2\pi$ will not introduce errors of larger order of magnitude than those already implicit in the linear equation of motion, provided that the body is *smooth* as well as slender.[5] Thus not only must the slope of the surface to the incident stream be small, but changes in it must be spread evenly over the length of the projectile, so that the curvature, too, is consistently small; and similarly the curvature must vary only gradually (actually, this condition is not essential if the pressure distribution is only required for calculating the drag). Under these rather stringent conditions of smoothness the boundary condition $v/(1 + u) = R'(x)$ on the surface $r = R(x)$, which may be approximated as $v = R'(x)$ since factors like $1 + u$ have already been neglected, becomes by a further approximation

$$f(x) = R(x)R'(x) = \frac{S'(x)}{2\pi} \tag{7-3}$$

This is because the expansion of v (see Eq. 7-1) for small r is

$$v = \frac{f(x)}{r} + \frac{1}{2}(M_\infty^2 - 1)r \int_0^x \left[\ln \frac{(M_\infty^2 - 1)^{\frac{1}{2}} r}{2(x - t)} - \frac{1}{2}\right] df''(t) + \cdots \tag{7-4}$$

which indicates that the approximation (7-3) will be tolerable if on the surface $r^2 f''(x)$ is small compared with $f(x)$, which is clearly the case under the conditions stated. (But notice that the error increases with Mach number.) The expansion (7-4) is most simply obtained from Heaviside's operational calculus, in which v is $pK_1(pr)f(x)$, if p is the operator

[5] The nomenclature here introduced, which divides the class of "slender" bodies according to whether or not they are "smooth," is more suggestive than the one common at present, which takes "slender" to imply "smooth."

$(M_\infty^2 - 1)^{\frac{1}{2}}\partial/\partial x$ and K_1 is a modified Bessel function of the second kind. Eq. 7-4 then corresponds to the expansion of the Bessel function for small pr [*44*, p. 80] with the operational forms interpreted.

It is important to notice that, since for smooth bodies the approximation (7-3) to the axial distribution of source strength $f(x)$ has a proportional error of order the *square* of the fineness ratio, this source strength is best taken as the value of $S'(x)/2\pi$ at the cross section *through* the source itself, *not* as the value of $S'(x)/2\pi$ at the cross section farther downstream where the influence of the source first emerges into the flow (which would clearly differ from it by a proportional error of order the first power of the fineness ratio).

By expanding u as v was expanded in Eq. 7-4, and using Eq. 7-2 and 7-3, we obtain the first order approximation to the pressure distribution on smooth slender pointed projectiles in axisymmetric flow in the form

$$C_p = \frac{1}{\pi} S''(x) \ln \frac{2x}{(M_\infty^2 - 1)^{\frac{1}{2}} R(x)} - \frac{1}{\pi} \int_0^x \frac{S''(x) - S''(t)}{x - t} dt - R'^2(x) \qquad (7\text{-}5)$$

The second and third terms are of order ϵ^2 (where ϵ is the fineness ratio). The mathematical order of magnitude of the first is strictly $\epsilon^2 \ln \epsilon^{-1}$; in practice, however, this is not very different from ϵ^2.

Flow past slender bodies which are not smooth. When, as in many practical problems, the projectile surface is not smooth, the function $f(t)$ must be found by some other means, of which the numerical solution of the boundary condition as an integral equation, which Kármán and Moore originally used [*41*], is a possibility. Another method is that introduced by the present author [*35*] in 1948. He showed that, as a result of a discontinuity in surface slope at $x = x_1$, where $R(x) = R_1$, by an amount $\Delta R'$, the pressure coefficient deduced from linear theory, with $f'(t) = S''(t)/2\pi$, and no account taken of the discontinuity, must be augmented by an amount

$$\frac{2(\Delta R')}{(M_\infty^2 - 1)^{\frac{1}{2}}} U\left(\frac{x - x_1}{(M_\infty^2 - 1)^{\frac{1}{2}} R_1}\right) \qquad (7\text{-}6)$$

where $U(z)$ is a function which is zero for $z < 0$, unity for $z = 0$, and falls thence asymptotically to zero as $z \to \infty$, as shown in Fig. E,7a (where the function z^{-1} is shown as a broken line for comparison). This fact enables any regions of large rate of change of slope or curvature to be treated by approximating to them by means of a number of small changes of slope. Indeed such an approach can be applied all along the surface, and the resulting sum allowed to approach, in the limit of a very large number of very small changes of slope, an integral. This gives

$$C_p = \frac{1}{\pi}\int_0^x U\left(\frac{x-t}{(M_\infty^2-1)^{\frac{1}{2}}R(t)}\right)\frac{dS'(t)}{(M_\infty^2-1)^{\frac{1}{2}}R(t)} - R'^2(x) \qquad (7\text{-}7)$$

as an approximate solution of Kármán's integral equation which is equally valid for any slender body, whether smooth or not. It is also correct for the external flow about any slender ducted body of revolution, provided that the integral is understood to include a contribution of the

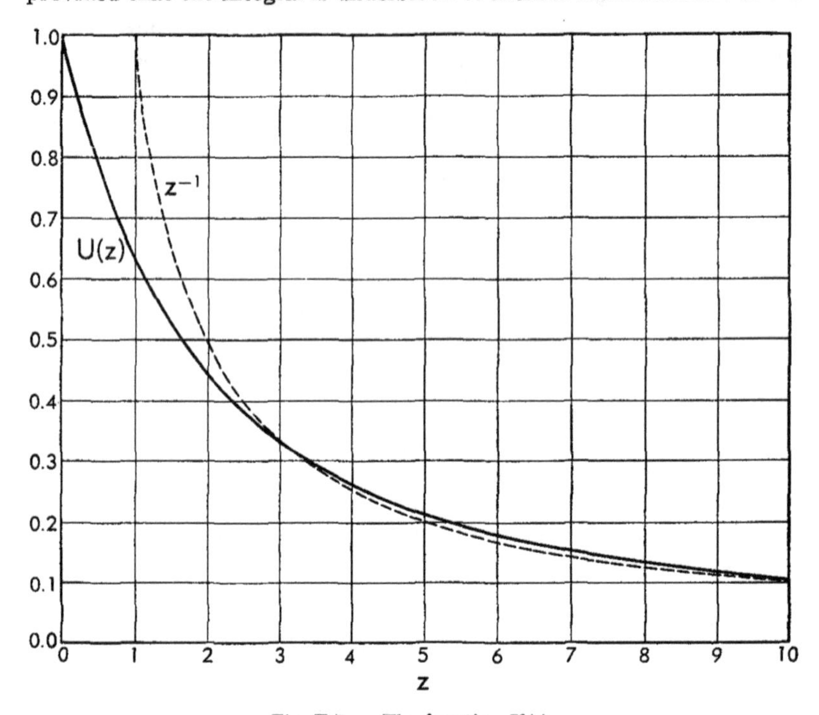

Fig. E,7a. The function $U(z)$.

form of Eq. 7-6, not only from places where the slope is discontinuous, but also from the lip, if there $S'(x) \neq 0$. This extreme versatility makes the formula (7-7) very useful. It is relatively easy to compute numerically, because the integrand is without singularities.

A brief explanation of the theory just stated will now be given. $U(z)$ is calculated as the function such that, if Eq. 7-1 are regarded as defining *for some constant value of* r (by elimination of f) u in terms of v, then

$$u(x, r) = -\int_0^x U\left(\frac{x-t}{(M_\infty^2-1)^{\frac{1}{2}}r}\right)\frac{dv(t, r)}{(M_\infty^2-1)^{\frac{1}{2}}} \qquad (7\text{-}8)$$

This function $U(z)$ was computed by the British Admiralty [*42*] by solving the integral equations when v is a unit function. But of course the integral equations (7-1) must in fact be solved not at constant r, but with r related to x by the equation $r = R(x)$ of the body surface. However, if the formula (7-8) be rewritten

$$u = -\int_0^x U\left(\frac{x-t}{(M_\infty^2-1)^{\frac{1}{2}}R(t)}\right)\frac{dS'(t)}{2\pi R(t)(M_\infty^2-1)^{\frac{1}{2}}} \tag{7-9}$$

two points are clear concerning the predicted contribution to u from any small change in rv ($= S'(t)/2\pi$ on the boundary): (1) It is approximately correct for small $x - t$, such that $R(x)$ has changed little from $R(t)$ (i.e. in this region r is approximately constant), by the definition of $U(z)$. (2) It is also approximately correct when $z = (x-t)/(M_\infty^2-1)^{\frac{1}{2}}R(t)$ is large enough for the approximation $U(z) \cong z^{-1}$ to be adequate (see Fig. E,7a for this), for then the addition to u due to the small change $dS'(t)$ has become, by Eq. 7-9,

$$-\frac{1}{x-t}\frac{dS'(t)}{2\pi} \tag{7-10}$$

This agrees with the first of Eq. 7-1 if $f(t)$ has adjusted itself fairly rapidly to the change in $S'(t)/2\pi$ at t, and that this must be so is clear from the second equation, as approximated, when $x - t$ is large compared with $(M_\infty^2-1)^{\frac{1}{2}}r$, by putting the change in f equal to the change in $rv = S'(x)/2\pi$.

Now these two regions will practically overlap for slender bodies.[6] Hence Eq. 7-9 is true not (as the author supposed [*42*] in 1945) only for ducted bodies of nearly constant cross section, but also for quite arbitrary slender bodies. Putting the reasoning of the last paragraph into words, this is because, though the pressure change due to a given change in slope is calculated by a formula deduced by assuming the cross-sectional area thereafter approximately constant, the said pressure change is actually *independent* of cross-sectional area sufficiently far downstream for the cross-sectional area to have altered significantly.

The formula (7-7) for C_p now follows at once from Eq. 7-9 and 7-2. The fact that for smooth bodies it agrees with Eq. 7-5, to within the order of approximation on which both are based, can be deduced without too much difficulty from the fact that

$$\int_0^x U(z)dz \sim \ln(2x) \text{ as } x \to \infty$$

Calculated examples are compared with experiment in Fig. E,7b. The upper figure represents the pressure distribution on a projectile consisting of a conical head of 10° semiangle, followed by a cylindrical

[6] Since the latter may be regarded as starting at $z = 2$, they overlap provided that $R(t + 2(M_\infty^2-1)^{\frac{1}{2}}R(t))$ is close to $R(t)$, which is true if $2(M_\infty^2-1)^{\frac{1}{2}}R'(t)$ is small. But even if they do not quite overlap, the approximation can hardly diverge far from the truth between them.

portion. In unpublished wind tunnel experiments by Cope at the NPL the measured pressures were clustered closely about the plain curve. The initial position of this curve is as calculated by Kopal [37] on exact adiabatic theory. The slight falls in pressure before the corner and

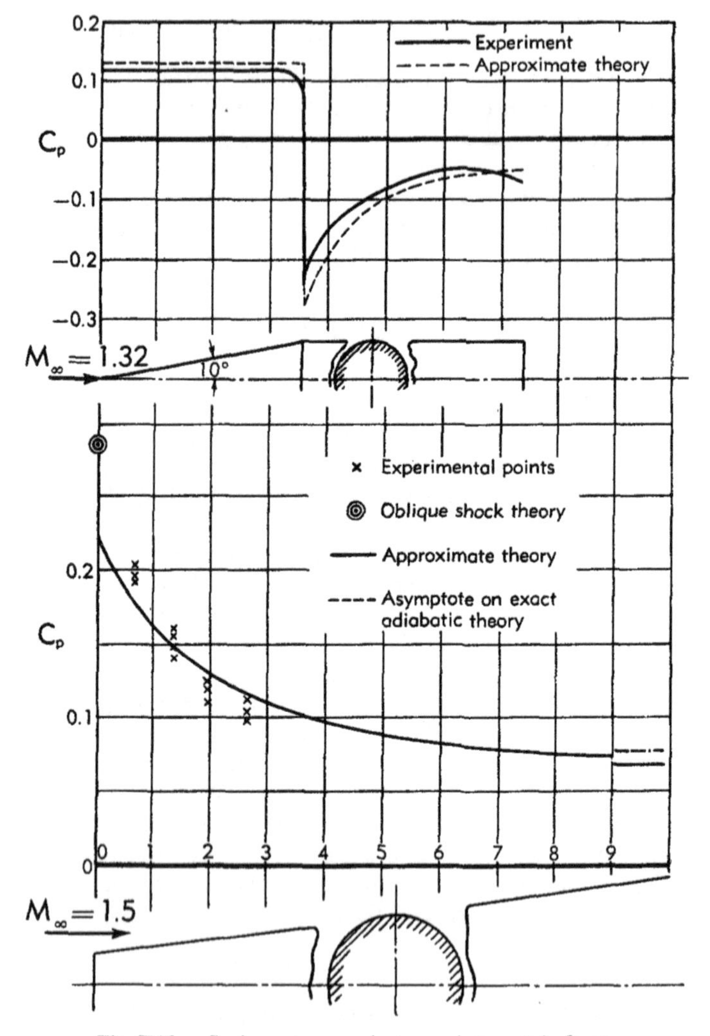

Fig. E,7b. Surface pressures in two axisymmetric flows.

before the base are boundary layer effects of familiar type. The discontinuity in the plain curve at the shoulder was calculated from the theory of the Prandtl-Meyer expansion, since the distance of the first experimental point behind it was as much as one third of the projectile radius. The dashed line is calculated on the author's theory and is in reasonable agreement. The lower figure represents the pressure distribution on the outside of a duct of conical shape, with semiangle 7.5°. The curve represents Eq. 7-7, calculated with $S(x) = \pi(1 + x \tan 7.5°)^2$ (measuring distances in multiples of the radius of the mouth). The value at the lip, $2 \tan 7.5°/\sqrt{M_\infty^2 - 1} - \tan^2 7.5°$, agrees to the first order with Ackeret's prediction of the pressure due to two-dimensional turning of the stream, but is of course an underestimate of the true pressure coefficient, which by oblique shock theory (Fig. E,3f) is as given by the double-ringed point. The crosses represent NPL measurements. The asymptote of the curve is represented by a plain line. The true asymptote of the pressure distribution, at least if boundary layer be neglected, is presumably the value for a solid cone, shown as a dashed line. In this lower figure the approximate theory is reasonably good except near the lip. Here the flow is nearly two-dimensional, which causes the disturbance pressure to be higher, and hence the error (of order its square) to be correspondingly greater. (The comparable, though smaller, error in the sudden fall of pressure in the upper figure is shared equally between the pressures ahead and behind, and so reduced in importance.) This shows how useful it would be if the theory could be improved so as to be uniformly correct to order ϵ^2, even in regions of pseudo-two-dimensional flow (where this would necessitate taking into account the nonlinearity of the equation of motion), so as to give (for example) the Busemann value (Art. 3) for the pressure change at a discontinuity of slope.

The author [*35*] was more concerned with total pressure drag than with pressure distribution. He showed that, of discontinuities, only those in slope affect the pressure drag significantly, and gave a fairly simple formula for evaluating it. (A discontinuity in *curvature* certainly alters the pressure distribution from that predicted by smooth body theory, but not so as to affect the drag to the order of approximation used.)

It should be emphasized in passing that no theory nearly so complete applies to the *internal* flow in slender ducts. Here, obviously, the linear theory applies only if the radius remains nearly constant; but even when this is so its predictions are difficult to interpret. Ward [*45*] showed that there are logarithmic infinities propagated along certain Mach lines in the linear theory, and it is still not understood what really happens in their neighborhood. (The theory in Lighthill's 1945 paper is altogether wrong.)

The theory of flow past bodies of revolution which are not smooth has been extended to a second approximation by Van Dyke [*54*], whose work therefore supersedes in point of accuracy Broderick's second order theory for smooth bodies [*34*], referred to below. Only a simplified account of Van Dyke's theory can be given here. If the perturbation potential is

$\varphi_1 + \varphi_2$ to a second approximation, where φ_1 is the first approximation, then φ_2 must be considered to satisfy not the ordinary Hantzsche-Wendt equation deduced from Eq. 2-14 but the more accurate equation

$$\frac{\partial^2\varphi_2}{\partial r^2} + \frac{1}{r}\frac{\partial\varphi_2}{\partial r} - (M_\infty^2 - 1)\frac{\partial^2\varphi_2}{\partial x^2}$$
$$= M_\infty^2 \frac{\partial}{\partial x}\left[\left(1 + \frac{\gamma - 1}{2} M_\infty^2\right)\left(\frac{\partial\varphi_1}{\partial x}\right)^2 + \left(\frac{\partial\varphi_1}{\partial r}\right)^2\right] + M_\infty^2\left(\frac{\partial\varphi_1}{\partial r}\right)^2\frac{\partial^2\varphi_1}{\partial r^2} \quad (7\text{-}11)$$

The cubic term on the extreme right has to be included because $\partial\varphi_1/\partial r$ and $\partial^2\varphi_1/\partial r^2$ are, as has been seen, considerably larger than $\partial\varphi_1/\partial x$ and $\partial^2\varphi_1/\partial x^2$ near the body.

Van Dyke's great advance, which rendered possible the solution of Eq. 7-11, was the discovery of a particular integral $\varphi_2 = \varphi_2^{(p)}$ of the equation. Here

$$\varphi_2^{(p)} = M_\infty^2\varphi_1\frac{\partial\varphi_1}{\partial x} + \frac{(\gamma + 1)M_\infty^4}{2(M_\infty^2 - 1)} r\frac{\partial\varphi_1}{\partial r}\frac{\partial\varphi_1}{\partial x} - \frac{1}{4}M_\infty^2 r\left(\frac{\partial\varphi_1}{\partial r}\right)^3 \quad (7\text{-}12)$$

The general solution takes the form of the sum of $\varphi_2^{(p)}$ with an arbitrary solution of the homogeneous equation which vanishes ahead of the undisturbed Mach cone. Thus

$$\varphi_2 = \varphi_2^{(p)} - \int_0^{x-(M_\infty^2-1)^{\frac{1}{2}}r} \frac{h'(t)dt}{[(x - t)^2 - (M_\infty^2 - 1)r^2]^{\frac{1}{2}}} \quad (7\text{-}13)$$

for some $h(t)$.

Van Dyke obtains φ_1 by solving the exact boundary condition numerically as an integral equation (following von Kármán and Moore [*41*]) and then solves numerically the boundary condition for φ_2 as an integral equation for $h(t)$. It is not possible to describe here his technique for numerical solution, which is rather complicated. But it should be mentioned that the divergences between theory and experiment observed in cases like those illustrated in Fig. E,7b (with the exception of those noted as boundary layer effects) are removed by the improvement in the theory. It is also particularly satisfactory that the theory continues to give excellent agreement with exact calculations (by the method of characteristics) even at a Mach number exceeding 3.

Nonaxisymmetric flows. To return to the questions (1) and (2) put at the beginning of this article, it is seen that the Van Dyke theory goes further than was suggested in those questions, which had already been answered in the pages preceding its mention. Two similar questions arise in the study of yawed flow past projectiles of revolution, and in Ward's theory [*43*] of flow past smooth slender bodies of more general cross section. In these cases, which will now be considered, a theory as complete as that of Van Dyke for axisymmetric flow does not at present exist. It is necessary to consider extensions of the preceding theories.

For the yawed flow the answer to both questions is obtainable along similar lines. Thus, for the theory with discontinuities in slope the reader is referred to the author's paper [*35*]; the over-all effect of these on the excess pressures due to yaw is found to be less than their effect on the pressures at zero yaw. Next, for smooth bodies at yaw the order of the errors can be deduced from his second approximation [*40*]. The problem is, however, a special case of Ward's more general theory [*43*], which applies to bodies (of any cross section) which are both smooth and slender in all directions perpendicular to the stream. (One can visualize how the extension of the Ward theory to general slender body shapes which are not smooth would be made, but it would probably be too complicated for practical use.) Ward gave a first order theory and also considered the order of magnitude of higher order terms.

A study of Broderick's second approximation [*34*] to axisymmetric flow, the author's work on smooth bodies at yaw [*40*] and Ward's estimate of second order terms in the general case [*43*] makes it clear that the true character of all these theories of flow over smooth bodies emerges clearly only from Ward's general approach. The first two methods perhaps attain their answers most expeditiously, but the work is still very long, and some of the answers are too complicated for extensive use. But the improvement of Ward's arguments, which will now be given, obtains in principle the second approximation for flow over an arbitrary smooth and slender body, and each stage in the argument illuminates the physics of the flow and the relative values of the approximations employed. Therefore the essential steps in the argument will here be carefully described. The results are not however calculated explicitly, but are taken direct from the above two papers [*34,40*], except that the most complicated term in Broderick's theory has been simplified in the light of Van Dyke's ideas.

The basis of the Ward theory is that, near the surface of a smooth slender pointed body, variation parallel to the axis is gradual compared with variation perpendicular to it (cf. C, 9 and D, 19–D, 24). Thus (working to the first order), if we write the linear equation for the disturbance potential φ, in cylindrical coordinates (r, θ, x) with the stream in the x direction (so that the full velocity potential is $q_\infty[x + \varphi]$), as

$$\frac{\partial^2\varphi}{\partial r^2} + \frac{1}{r}\frac{\partial\varphi}{\partial r} + \frac{1}{r^2}\frac{\partial^2\varphi}{\partial\theta^2} = (M_\infty^2 - 1)\frac{\partial^2\varphi}{\partial x^2} \tag{7-14}$$

then near the body the left-hand side of the equation is dominant, and effectively the solution near the body is a plane harmonic in any plane perpendicular to the stream. The boundary condition is obvious from Fig. E,7c, in which two cross sections (normal to the stream) at stations a distance dx apart are shown superposed. Here dn, at any point on either, is the perpendicular distance between them. Expressing that there is no flow component normal to the surface, we see that in either plane section

$$\frac{\partial \varphi}{\partial n} = \left(1 + \frac{\partial \varphi}{\partial x}\right)\frac{dn}{dx} \tag{7-15}$$

which to a linear approximation becomes

$$\frac{\partial \varphi}{\partial n} = \frac{dn}{dx} \tag{7-16}$$

However, the determination of the appropriate plane harmonic describing the flow in a plane section near the body surface, and satisfying

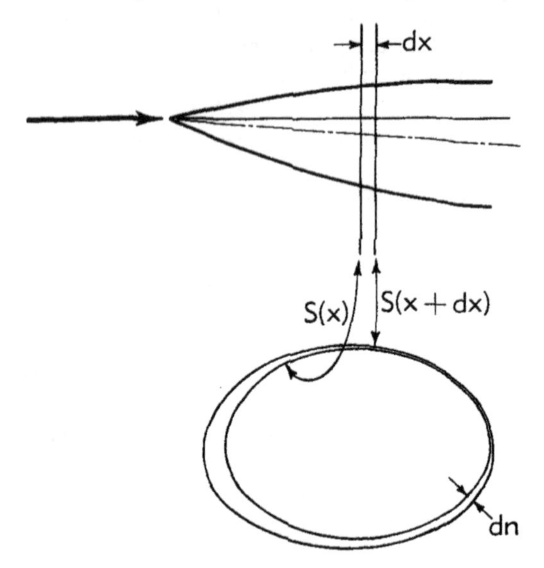

Fig. E,7c. Two superimposed adjacent cross sections of a smooth slender body.

the boundary condition (7-16), is not unique. It is obvious, for example, that any "constant" term (i.e. one depending on x alone) may be added without disturbing the conditions. Nor is it clear that the harmonic must be one which farther from the body has vanishingly small gradients, because, of course, the harmonic does not actually represent the potential out there, where the right-hand side of Eq. 7-14 is of equal importance with the left-hand side.

In fact the solution can be rendered unique only by continuing it away from the body surface into a solution of the full equation (7-14), and expressing the condition that the disturbance potential so obtained vanishes upstream of the Mach cone from the nose. But luckily this process can be done once for all in the general case. For if one considers the potential expanded in a Fourier series in θ, away from the axis, then the term independent of θ must be given by the axial source distribution potential

$$-\int_0^{x-(M_\infty^2-1)^{\frac{1}{2}}r} \frac{f(t)dt}{[(x-t)^2-(M_\infty^2-1)^2r^2]^{\frac{1}{2}}} \tag{7-17}$$

(whose derivatives with respect to x and r were given in Eq. 7-1), and the other terms are given by a similar axial distribution of multipoles, such as

$$\cos n\theta \int_0^{x-(M_\infty^2-1)^{\frac{1}{2}}r} \frac{F_n(t)dt}{[(x-t)^2-(M_\infty^2-1)r^2]^{\frac{1}{2}}} \left(\left\{\frac{x-t+[(x-t)^2-(M_\infty^2-1)r^2]^{\frac{1}{2}}}{(M_\infty^2-1)^{\frac{1}{2}}r}\right\}^n + \left\{\frac{(M_\infty^2-1)^{\frac{1}{2}}r}{x-t+[(x-t)^2-(M_\infty^2-1)r^2]^{\frac{1}{2}}}\right\}^n\right) \tag{7-18}$$

(These are obtained most simply by Heaviside's operational theory; for example Eq. 7-18 is $2K_n(pr)F_n(x)\cos n\theta$, where, as above, p stands for $(M_\infty^2-1)^{\frac{1}{2}}\partial/\partial x$; this satisfies Eq. 7-14; it also represents a potential zero ahead of the Mach cone through the nose, because the function $K_n(z)$ is like $e^{-z}\sqrt{\pi/2z}$ for large z, and, for example, $e^{-pr}F_n(x)$ signifies $F_n(x-(M_\infty^2-1)^{\frac{1}{2}}r)$.) Now for small r (i.e. in the region near the body, where the potential is expected to take the form of a plane harmonic) Eq. 7-17 becomes approximately

$$f(x)\ln r + \int_0^x \ln\left[\frac{(M_\infty^2-1)^{\frac{1}{2}}}{2(x-t)}\right]f'(t)dt \tag{7-19}$$

(this again is best obtained by expanding $-K_0(pr)f(x)$ for small r) and Eq. 7-18 becomes approximately

$$\frac{2^n}{(M_\infty^2-1)^{\frac{1}{2}n}}\frac{\cos n\theta}{r^n}\int_0^x (x-t)^{n-1}F_n(t)dt \tag{7-20}$$

Hence the whole potential near the surface is a combination of harmonics like Eq. 7-20, which vanish at infinity, with a part (7-19) involving a constant term and a term proportional to $\ln r$. Further these last two terms are *related;* the constant term is expressible in terms of the coefficient of $\ln r$.

Since, then, the plane harmonic required to satisfy the boundary condition (7-16) has zero gradient at infinity, it is specified uniquely by this boundary condition except for the possible addition of a function of x only. But the latter function is specified in terms of the coefficient of $\ln r$, as has just been seen. Also the coefficient of $\ln r$, by Gauss's theorem, is

$$\frac{1}{2\pi}\oint\frac{\partial\varphi}{\partial n}ds = \frac{1}{2\pi}\oint\frac{dn}{dx}ds = \frac{S'(x)}{2\pi} \tag{7-21}$$

Here the integral is taken around the cross section, and the geometry of Fig. E,7c makes it clear that the second integral is equal (as stated) to the rate of change $S'(x)$ of cross-sectional area (normal to the stream) with x. Hence the true form of the axisymmetric part of the first order disturbance potential (i.e. the term independent of θ in its Fourier analysis), by Eq. 7-19, is

$$\frac{1}{2\pi}\int_0^x \ln\frac{(M_\infty^2-1)^{\frac{1}{2}}r}{2(x-t)}S''(t)dt \tag{7-22}$$

This result for the axisymmetric part of the potential is thus independent of whether the projectile is one of revolution.

It should be added that if the boundary condition (7-16) is to be satisfied at the surface, by a sum of Eq. 7-22 with terms like Eq. 7-20, it must doubtless be roughly the case that the terms are of the same order of magnitude for values of r comparable with the values at the body surface. Thus the coefficient $F_n(t)$ in Eq. 7-18 must be smaller than that of Eq. 7-17, i.e. $f(t) = S'(t)/2\pi$, by an order of magnitude the nth power of the body radius measured from the axis. This proves what was stated in Art. 6, that, for given x, at distances from the axis large compared with the body radius for that x, these nonaxisymmetric terms are relatively unimportant.

Nonaxisymmetric flows to a second approximation. Having stated Ward's linear theory (for applications of which see [*43*]), we now investigate the second approximation to the flow. Near the body surface this is governed by an equation of which the left-hand side, embodying the largest terms, is as in Eq. 7-14, and of which the right-hand side contains those terms (out of the full equation of potential flow) next in importance; these include, besides the linear term in Eq. 7-14, nonlinear terms. The said equation is, in fact,

$$\frac{\partial^2\varphi}{\partial r^2}+\frac{1}{r}\frac{\partial\varphi}{\partial r}+\frac{1}{r^2}\frac{\partial^2\varphi}{\partial\theta^2} = -\frac{\partial^2\varphi}{\partial x^2}$$
$$+\frac{1}{2}M_\infty^2\left(\frac{\partial}{\partial x}+\frac{\partial\varphi}{\partial r}\frac{\partial}{\partial r}+\frac{1}{r^2}\frac{\partial\varphi}{\partial\theta}\frac{\partial}{\partial\theta}\right)\left[2\frac{\partial\varphi}{\partial x}+\left(\frac{\partial\varphi}{\partial r}\right)^2+\frac{1}{r^2}\left(\frac{\partial\varphi}{\partial\theta}\right)^2\right] \tag{7-23}$$

The selection from the full equation of the terms retained in Eq. 7-23 is based on the principle that axial variation (represented by $\partial/\partial x$) should be slower than normal variation ($\partial/\partial r$ and $r^{-1}\partial/\partial\theta$) by a factor of the order of the normal disturbance velocities $\partial\varphi/\partial r$ and $r^{-1}\partial\varphi/\partial\theta$. In particular squares of these latter velocities are comparable with $\partial\varphi/\partial x$. Notice that the convection of sound has to be rather fully taken into account. On the other hand, terms due to variation in the speed of sound are negligible compared with those retained in Eq. 7-23, a typical such term being $(\gamma-1)M_\infty^4(\partial\varphi/\partial x)(\partial^2\varphi/\partial x^2)$.

To obtain the second approximation $\varphi = \varphi_1 + \varphi_2$, where φ_1 is the first approximation near the body as calculated by Ward (roughly, φ_1 is of the order of the square of the fineness ratio, and φ_2 may be expected to be of the order of its fourth power) we may solve Eq. 7-23 with φ replaced on the right-hand side by φ_1. A particular integral is found fairly easily by complex-variable methods. If $\zeta = re^{i\theta}$, then φ_1, as a plane harmonic, is the real part of a regular function $w(\zeta)$. Eq. 7-23 may be written, using this, as

$$\frac{\partial^2\varphi}{\partial\zeta\partial\bar{\zeta}} = \frac{1}{8}(M_\infty^2 - 1)\frac{\partial^2}{\partial x^2}(w + \bar{w}) + \frac{1}{8}M_\infty^2\left[2\frac{dw}{d\zeta}\frac{d}{d\bar{\zeta}}\left(\frac{\partial\bar{w}}{\partial x}\right)\right.$$
$$\left. + 2\frac{d\bar{w}}{d\bar{\zeta}}\frac{d}{d\zeta}\left(\frac{\partial w}{\partial x}\right) + \left(\frac{dw}{d\zeta}\right)^2\frac{d^2\bar{w}}{d\bar{\zeta}^2} + \left(\frac{d\bar{w}}{d\bar{\zeta}}\right)^2\frac{d^2w}{d\zeta^2}\right] \tag{7-24}$$

of which a particular integral is the real part of

$$\frac{1}{4}(M_\infty^2 - 1)\bar{\zeta}\int\frac{\partial^2 w}{\partial x^2}d\zeta + \frac{1}{2}M_\infty^2 w\frac{\partial\bar{w}}{\partial x} + \frac{1}{4}M_\infty^2\frac{d\bar{w}}{d\bar{\zeta}}\int\left(\frac{dw}{d\zeta}\right)^2 d\zeta \tag{7-25}$$

(The physical significance of the terms in Eq. 7-25 will be discussed later.)

To this particular integral must be added a complementary function, i.e. a solution of Eq. 7-23 with right-hand side replaced by zero, in other words, a plane harmonic. This must be chosen to make $\varphi_1 + \varphi_2$ a one-valued function of r and θ satisfying, to within the approximation employed, the more exact boundary condition (7-15). Since φ_1 already satisfies Eq. 7-16, we must have in fact

$$\frac{\partial\varphi_2}{\partial n} = \frac{\partial\varphi_1}{\partial x}\frac{dn}{dx} \tag{7-26}$$

But it is clear that, again, φ_2 is so far indeterminate, at least to the extent of an arbitrary function of x alone. For Eq. 7-26 concerns only the derivatives $\partial\varphi_2/\partial r$ and $r^{-1}\partial\varphi_2/\partial\theta$. The specification of the part of φ_2 independent of r and θ can only be obtained, as was that of φ_1, by extending the solution outside the region near the surface, this time to a second approximation, and expressing the condition that the extended value of φ_2 vanishes ahead of the Mach cone from the nose. This process ultimately yields a relation between the coefficients of $\ln r$ and 1 in φ_2, whence the latter coefficient is determined.

Broderick carried out this process [*34*] for axisymmetric flow, and actually the result must be precisely the same in the general case of Ward's slender bodies as can be shown from the fact that, away from the axis, the flow *is* axisymmetric to a first approximation; but this part of his work exceeds by a large factor, both in length and in complexity of the final answer, all the rest of the present investigation. Now, however, it can be greatly shortened in both respects, by use of Van Dyke's form (7-13) of second approximation to the potential. For smooth bodies the form of this near the body is obtained by replacing φ_1 by Eq. 7-22 and expanding the integral as in Eq. 7-19; it is

$$M_\infty^2 \left[f(x) \ln r + g(x) + \frac{(\gamma+1)M_\infty^2}{2(M_\infty^2 - 1)} f(x) \right] [f'(x) \ln r + g'(x)]$$

$$- \frac{1}{4} M_\infty^2 \frac{f^3(x)}{r^2} + h(x) \ln r + \int_0^x \ln \left[\frac{(M_\infty^2 - 1)^{\frac{1}{2}}}{2(x-t)} \right] h'(t) dt \quad (7\text{-}27)$$

Here $f(x) = S'(x)/2\pi$, as always in smooth body theory, and Eq. 7-22 has been written $f(x) \ln r + g(x)$. Of course the second term φ_2 in the approximation for smooth bodies must contain not only Eq. 7-27 but also the second term in the expansion near the body of the full linearized solution (7-17). However, this does not affect the coefficients of ln r and 1, which must be related as in Eq. 7-27. Thus, if

$$M_\infty^2 (fg)' + \frac{(\gamma+1)M_\infty^4}{2(M_\infty^2 - 1)} ff' \quad \text{and} \quad M^2 gg' + \frac{(\gamma+1)M_\infty^4}{2(M_\infty^2 - 1)} fg' \quad (7\text{-}28)$$

respectively be subtracted from these coefficients, the remainders are related as are the coefficients of ln r and 1 in Eq. 7-19.

This relationship is still complicated, but Broderick's [*34*] was so much more so that it could not be evaluated except for the single case of flow past a cone (see below). However, it should be noted that the complexity of the term independent of r and θ in φ_2 does not affect all aspects of the investigation, since for many purposes (e.g. evaluating lifts and moments) only the other terms in φ_2 are needed.

There is one other indeterminacy in φ_2, that is to say one other plane harmonic, with zero normal derivative on the boundary, of which any multiple could (without violating the conditions on the flow near the body) be added to φ_2, so that the particular multiple needs to be determined by a special device. This is the harmonic function which behaves far from the body like $r \cos(\theta + \delta)$, for some δ, and so represents the two-dimensional incompressible flow of a uniform stream past the cross section. But the determination of the appropriate multiple of *this* for incorporation in φ_2 is no more difficult than Ward's determination of the term independent of r and θ in φ_1. This is because, if a term of order $E(x) r \cos(\theta + \delta)$ is to be significant, even to the second approximation, near the body, then $E(x)$ must be of the order of the cube of the fineness ratio, and this (considered *not* near the body) is of the order of the coefficients of $\cos(\theta + \delta)$ in the *first* approximation φ_1. Hence $E(x)$ may be studied simply on the linear theory. The solutions of the linear equation of motion which are proportional to $\cos(\theta + \delta)$, and vanish ahead of the Mach cone from the nose, are of the form

$$\frac{\cos(\theta + \delta)}{r} \int_0^{x - (M_\infty^2 - 1)^{\frac{1}{2}} r} \frac{(x - t) B'(t) dt}{[(x-t)^2 - (M_\infty^2 - 1) r^2]^{\frac{1}{2}}} \quad (7\text{-}29)$$

for some $B(x)$. But the expansion of Eq. 7-29 near the body is (cf. Eq. 7-4)

$$\cos(\theta+\delta)\left\{\frac{B(x)}{r} + \frac{1}{2}(M_\infty^2-1)r\int_0^x\left[\ln\frac{(M_\infty^2-1)^{\frac{1}{2}}r}{2(x-t)}-\frac{1}{2}\right]dB''(t)+\cdots\right\} \qquad (7\text{-}30)$$

which shows that the coefficient $E(x)$ of $r\cos(\theta+\delta)$ in φ_2 is obtained at once from that of $r^{-1}\cos(\theta+\delta)$ in φ_1 (which has here been called $B(x)$) in the form

$$E(x)=\frac{1}{2}(M_\infty^2-1)\int_0^x\left[\ln\frac{(M_\infty^2-1)^{\frac{1}{2}}}{2(x-t)}-\frac{1}{2}\right]dB''(y) \qquad (7\text{-}31)$$

On the other hand the fact that, by expression (7-30), the coefficient of

$$(r\ln r)\cos(\theta+\delta) \qquad (7\text{-}32)$$

in φ_2 must be $\frac{1}{2}(M_\infty^2-1)B''(x)$, is merely an independent check on the work; indeed it is clear that the part of the first term in Eq. 7-25 corresponding to the term $B(x)e^{-i\delta}\zeta^{-1}$ in $w(\zeta)$ will, when a corresponding term with $\zeta e^{2i\delta}$ for $\bar{\zeta}$ has been added to it to make it single-valued, have just this behavior for large r.

There are no further indeterminacies in φ_2 to be rendered determinate by such devices. For the only single-valued harmonic functions with zero normal derivative on the boundary are those behaving at large distances like $r^n\cos(n\theta+\delta)$, for some δ, where n is a non-negative integer. Such a function could have a place in φ_2 only if multiplied by a coefficient of order the $(4-n)$th power of the fineness ratio. But Ward's theory shows (see the paragraph following Eq. 7-22 above) that even the first approximation φ_1 has, as coefficient of $\cos(n\theta+\delta)$ when r is not small, a term at most of order the $(2+n)$th power of the fineness ratio. Hence such a harmonic occurs in φ_2 only if $2+n\leqq 4-n$ or $n\leqq 1$. (If there were any question of going to still higher approximations to φ, which there is not, the said function would occur from φ_{n+1} onward.)

The whole process for determining φ_2 having now been specified, we may draw general conclusions from its form. Conclusions regarding its order of magnitude (for use in a critique of the first order theory) may reasonably be drawn from Eq. 7-25, since the operations to be performed on its various terms to obtain φ_2 should not greatly change their order of magnitude, and since the additional term $E(x)r\cos(\theta+\delta)$ is, by Eq. 7-31, of similar order to the first term in Eq. 7-25. Regarding the *relative* orders of magnitude of the three terms in Eq. 7-25, observe that the last two, which represent corrections to the linear theory due to the convection of sound, are for smooth bodies just as important as the first term, which represents corrections due to departures, near the body, of the linear equation of motion (7-14) from Ward's simplified form neglecting the right-hand side. Thus *for smooth bodies, if the linear equation of motion be used at all, the full smooth body approximation can be used without serious additional loss of accuracy.* (Further numerical evidence for this is given below, Fig. E,7d.)

Eq. 7-25 shows clearly that this conclusion is independent of Mach number. The error resulting from both approximations varies as M_∞^2. This is because, on the one hand, variation parallel to the axis is less gradual for larger M_∞, since the Mach lines make a smaller angle with the stream, and, on the other hand, the variations in fluid velocity become a larger fraction of the speed of sound as M_∞ increases.

The actual orders of magnitude of the three terms relative to φ_1 itself are

$$M_\infty^2 \frac{r^2}{x^2}, \qquad M_\infty^2 \frac{\varphi_1}{x}, \qquad M_\infty^2 \frac{\varphi_1^2}{r^2} \tag{7-33}$$

if smoothness is taken to mean that $\partial/\partial x$ is of order x^{-1}. Since φ_1/x is of the order of the square of the fineness ratio ϵ, it is seen that roughly the error due to each term (near the body) represents a factor $M_\infty^2\epsilon^2$. It is clear from this that the linear theory can be used, for practical body shapes, only at moderate supersonic speeds. This conclusion has been drawn by several writers, from different kinds of argument.

Broderick's theories and their significance. The specialization of the above theory to the axisymmetric and yawed flows past projectiles of revolution will now be indicated very briefly indeed, and the results discussed. In the axisymmetric case φ_1 consists only of the term (7-19) independent of θ, so $w(\zeta)$ is of the form $f(x) \ln \zeta + g(x)$, where as always in smooth body theory $f(x) = S'(x)/2\pi$, while $g(x)$ stands for expression (7-22) with r omitted. Hence the real part of Eq. 7-25 is easily calculated as a sum of multiples of $r^2 \ln r$, r^2, $\ln^2 r$, $\ln r$, 1, and r^{-2} by specific functions of x. (At least this is so after a term in θ^2 has been removed, by subtracting a suitable multiple of the real part of $\ln^2 \zeta$, in order to make the result single-valued.) To satisfy the boundary condition (7-26) on $r = R(x)$ a new multiple of $\ln r$ has to be added. The terms in φ_2 other than the term independent of r are thus rapidly calculated as

$$\begin{aligned} \tfrac{1}{4}(M_\infty^2 - 1)[f''r^2 \ln r + (g'' - f'')r^2] + M_\infty^2 ff' \ln^2 r + \{(f' \ln R + g')RR' \\ - \tfrac{1}{4}(M_\infty^2 - 1)(2f'' \ln R + 2g'' - f'')R^2 - 2M_\infty^2 ff' \ln R \\ - \tfrac{1}{2}M_\infty^2 f^3 R^{-2}\} \ln r - \tfrac{1}{4}M_\infty^2 f^3 r^{-2} \end{aligned} \tag{7-34}$$

The remaining term is then given by the rule incorporating expressions (7-28).

Broderick's method [34] of obtaining these terms was conceptually different. He considered the potential everywhere as expanded in powers of the fineness ratio ϵ, for a series of body shapes $r = \epsilon R(x)$. He had to consider the coefficients of ϵ^2, ϵ^4, and ϵ^6 in this expansion, and also, as the terms in $\ln R$ in Eq. 7-34 indicate, the coefficient of $\epsilon^4 \ln \epsilon^{-1}$. He pointed out that terms in $\epsilon^6 \ln \epsilon^{-1}$ and $\epsilon^6 \ln^2 \epsilon^{-1}$ exist, and also investigated terms in still higher powers of ϵ. He expanded the equation for each of these coefficients in ascending powers of r and $\ln r$, to determine their form for small r. Finally he picked out the terms which are of order ϵ^2 or $\epsilon^2 \ln \epsilon^{-1}$ near the body, namely, Eq. 7-22 (the first approximation for smooth slender bodies), and then the terms which are of order ϵ^4 or $\epsilon^4 \ln \epsilon^{-1}$ or $\epsilon^4 \ln^2 \epsilon^{-1}$ near the body, namely (apart from the term independent of r), Eq. 7-34. He had also to prove that no contribution arises from coefficients of other powers of ϵ and $\ln \epsilon^{-1}$. There is little doubt that the present method, in which the equation for the potential is approximated in a different manner, appropriate to flow conditions near the body, and the relevant terms emerge as the whole of the first and second approximations respectively, is superior for obtaining Eq. 7-34.

The pressure coefficient is calculated to the order of approximation required as

$$C_p = -2\frac{\partial\varphi_1}{\partial x} - \left(\frac{\partial\varphi_1}{\partial r}\right)^2 - \left(\frac{\partial\varphi_1}{\partial x}\right)^2 - 2\frac{\partial\varphi_2}{\partial x} - 2\frac{\partial\varphi_1}{\partial r}\frac{\partial\varphi_2}{\partial r} + \frac{1}{4}M_\infty^2\left[2\frac{\partial\varphi_1}{\partial x} + \left(\frac{\partial\varphi_1}{\partial r}\right)^2\right] \tag{7-35}$$

of which the first two terms constitute the first approximation and are of orders $\epsilon^2 \ln \epsilon^{-1}$ and ϵ^2 near the body, while the remaining terms are of orders $\epsilon^4 \ln^2 \epsilon^{-1}$, $\epsilon^4 \ln \epsilon^{-1}$, and ϵ^4 near the body. They are easily calculable from Eq. 7-22 and 7-34, except for the term depending on x alone arising from the fourth term of Eq. 7-35, which arises from that in φ_2. It is not here given, but it may be mentioned that it is of order ϵ^4.

For the flow past a cone a numerical comparison of the various terms is possible. If ϵ is the tangent of the semivertical angle, Eq. 7-35 becomes

$$C_p = -\epsilon^2 + 2\epsilon^2 \ln\frac{2}{(M_\infty^2-1)^{\frac{1}{2}}\epsilon} + 3(M_\infty^2-1)\epsilon^4\left[\ln\frac{2}{(M_\infty^2-1)^{\frac{1}{2}}\epsilon}\right]^2 - (5M_\infty^2-1)\epsilon^4 \ln\frac{2}{(M_\infty^2-1)^{\frac{1}{2}}\epsilon} + \left[\frac{13}{4}M_\infty^2 + \frac{1}{2} + \frac{(\gamma+1)M_\infty^4}{M_\infty^2-1}\right]\epsilon^4 \tag{7-36}$$

of which the first two terms constitute the first order approximation. The term involving γ comes from Broderick's evaluation of the term independent of r in φ_2. In Fig. E,7d the pressure coefficient on a cone is plotted against Mach number for semiangles of 10° and 15°. Here the full lines represent Kopal's numerical calculation [*37*] of the exact adiabatic solution, the dashed lines represent the second order approximation (7-36), the dotted lines represent the first order approximation (the first

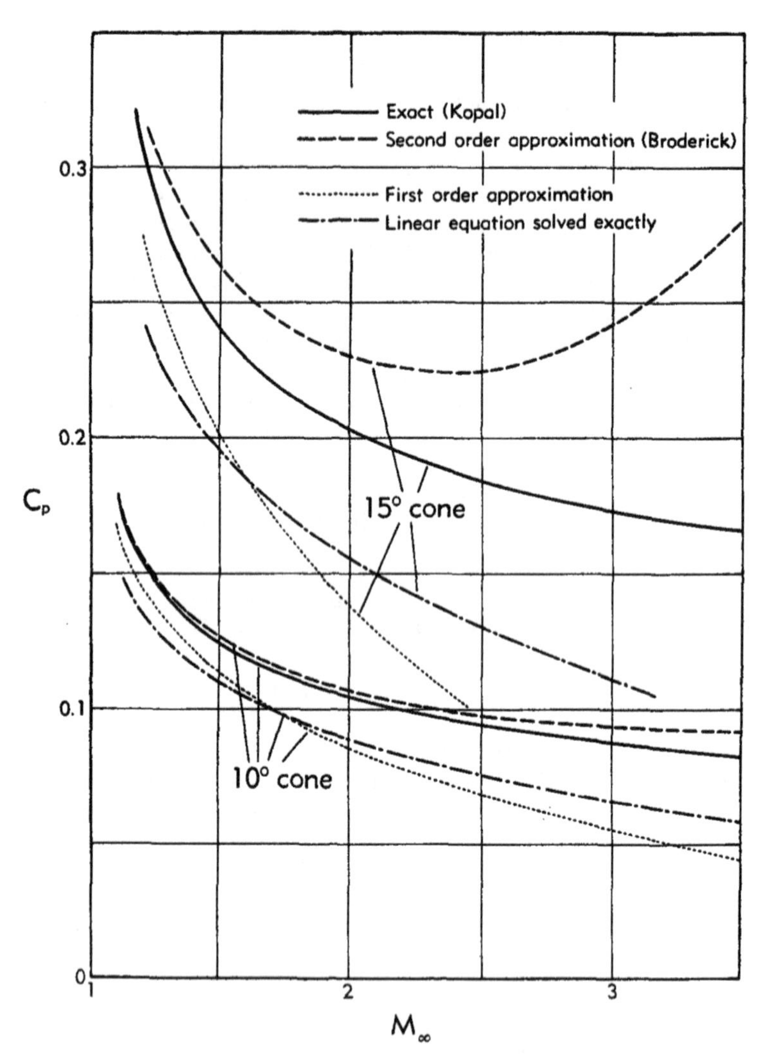

Fig. E,7d. Pressure coefficient on the surface of a cone in axisymmetric flow at Mach number M_∞.

two terms of Eq. 7-36), while the dash-dotted line is the curve obtained by using the linearized equation for the velocity field and satisfying the boundary condition at the surface exactly, and also deducing the pressure coefficient from the exact form of Bernoulli's equation. It is seen that, as predicted above from the orders of magnitude of the terms, there is, in the case of a cone, hardly any appreciable advantage, if the linear equation is being used at all, in not adopting the full smooth body approximation. For a semiangle of 5° the exact and Broderick's curves are indistinguishable in the range of M_∞ used in Fig. E,7d. It is seen that the Broderick theory is still excellent for 10°, but is beginning to be a serious overestimate for 15°. The errors in all the theories increase markedly with Mach number, as predicted above. (They are really all meaningless for semiangle 15° and $M_\infty = \csc 15° = 3.86$, since the cone then coincides with the undisturbed Mach cone.)

In Broderick's second paper in the journal cited *[34]* he extended further his solution for the flow past a cone to include the flow near the shock. In this way he was able to prove rigorously that his value (7-36) for the surface pressure was correct if terms of order $\epsilon^6 \ln^3 \epsilon^{-1}$ were neglected, even if full account be taken of the properties of the shock. This was hardly obvious (except by analogy with the two-dimensional flow) since the second approximation to the disturbance potential is of the same order of magnitude as the shock strength. Broderick's proof, therefore, sets the whole theory (not simply in the special case of the cone) on a firmer footing. To be precise, he showed that his series for φ in powers of ϵ converges only in a certain region stopping short of the undisturbed Mach cone through the nose. But he obtained a second series expansion in powers of ϵ whose coefficients are functions of the ratio of distance from the shock to distance of the Mach cone from the shock. This expansion converged in a region including the shock and overlapping the former region, whereas it satisfied the full Rankine-Hugoniot condition at the shock. By comparison of the series in their common region of convergence he was able to determine all remaining arbitrary constants that were needed to obtain the full second approximation to the surface pressures.

Broderick's detailed conclusions regarding flow near the shock are easily understood in the light of Whitham's subsequent theory (Art. 6). He finds that in this region $v = \partial\varphi/\partial r$ (for example) is given, to a first approximation, in the notation of Art. 6, by

$$v \cong 2^{\frac{1}{2}}(M_\infty^2 - 1)^{\frac{1}{4}}\delta^2 \left\{\left[\frac{x}{r} - (M_\infty^2 - 1)^{\frac{1}{2}} + k^2\delta^4\right]^{\frac{1}{2}} + k\delta^2\right\} \qquad (7\text{-}37)$$

having obtained certain of the constants herein by a comparison (involving careful consideration of error terms) with the small perturbation theory solution

$$v \cong 2^{\frac{1}{2}}(M_\infty^2 - 1)^{\frac{1}{4}}\delta^2 \left[\frac{x}{r} - (M_\infty^2 - 1)^{\frac{1}{2}}\right]^{\frac{1}{2}} \qquad (7\text{-}38)$$

On the other hand Eq. 7-37 agrees perfectly with Whitham's (6-13) if $F(y)$ is given by Eq. 6-12 and if this value is also used to solve Eq. 6-6 as a quadratic in $y^{\frac{1}{2}}$. The result, like Eq. 6-12, must be true all along the shock for the cone, and also, it may be mentioned, along the extensive initial stretch of it for any projectile.

But there is yet another interpretation of Eq. 7-37, which is important in the general theory (Art. 8) of the shock, and of the second approximation to the surface pressures, in "conical flows." In this interpretation, the first term in braces is regarded as identical with the linear theory solution (7-38), except that the value of x/r is modified to become $(x/r) + k^2\delta^4$, so that, as it were, the "undisturbed Mach cone" (where the solution is singular) is shifted forward, and indeed *beyond* where the solution is cut off by the front shock. This "doctoring" of the linear solution is correct (as Eq. 7-37 shows) near the shock, while elsewhere, where $x/r - (M_\infty^2 - 1)^{\frac{1}{2}}$ is not small, the error is of order δ^6 and so negligible even to a second approximation. On the other hand the second term in braces in Eq. 7-37 is regarded as due to the second approximation φ_2, which has

$$\frac{\partial \varphi_2}{\partial r} = 2^{\frac{1}{2}}(M_\infty^2 - 1)^{\frac{1}{4}}k\delta^4 \tag{7-39}$$

on the undisturbed Mach cone $x/r = (M_\infty^2 - 1)^{\frac{1}{2}}$. Eq. 7-39 is in fact true, trivially, for the "natural" second approximate solution, whose equation away from the body surface is not Eq. 7-23 or 7-11 but follows from Eq. 2-14 in the form

$$\frac{\partial^2 \varphi_2}{\partial r^2} + \frac{1}{r}\frac{\partial \varphi_2}{\partial r} - (M_\infty^2 - 1)\frac{\partial^2 \varphi_2}{\partial x^2}$$
$$= \frac{\partial}{\partial x}\left[\left(1 + \frac{\gamma - 1}{2}M_\infty^2\right)\left(\frac{\partial \varphi_1}{\partial x}\right)^2 + \left(\frac{\partial \varphi_1}{\partial r}\right)^2\right] \tag{7-40}$$

If φ_2 is taken as zero (because assumed continuous) on the Mach cone $x/r = (M_\infty^2 - 1)^{\frac{1}{2}}$, the first and third terms on the left vanish thereon; the right-hand side is easily calculated as r^{-1} times that of Eq. 7-39, whose truth therefore follows. It is a good approximation not only on the Mach cone but also anywhere sufficiently near to it, and hence Eq. 7-37 follows.

This gives a new principle (analogous to but essentially different from Whitham's of Art. 6) for obtaining a flow field which is everywhere correct to a second approximation, from knowledge of the second approximate flow field calculated by the straightforward process of successive approximation. The principle is as follows: *The second approximate flow field is uniformly valid if distance downstream from the undisturbed Mach cone is always interpreted therein as distance downstream from the limit cone.*

In the case of flow past a cone the limit cone referred to is given by $x/r = (M_\infty^2 - 1)^{\frac{1}{2}} - k^2\delta^4$. We are justified in calling this cone, on which the flow field is singular, a limit cone (by analogy with limit line in two dimensions) because it is in fact the envelope of the Mach lines (6-6), as the ordinary process of differentiation with respect to y and elimination of y shows.

This principle could probably, in theory, be applied to determine the shocks on general (nonconical) projectiles, but it is very unsatisfactory for this purpose (compared with Whitham's principle), since it involves considering the second approximation. However, it is the principle which must be applied in the general theory of conical fields (Art. 8), and so it is important to notice that in general physical soundness and plausibility it is the equal of Whitham's. It states in fact that the square root singularities in the first and second approximate solutions correspond to a real singularity in the theoretical potential flow, i.e. a limit line; notice that near the limit line the velocities necessarily vary like the square root of distance from it. But this singularity is really slightly ahead of where it is predicted to be on these approximate solutions, and therefore the flow field must be shifted forward so that the predicted singularity comes in the right place. This makes a big difference to the behavior in the neighborhood of the shock, but elsewhere the shift in the coordinates by a multiple of δ^4 produces a negligible effect to the order of approximation used. The shock has to be determined as a surface between the limit cone and the undisturbed Mach cone, such that the deflection behind the shock (predicted on the above principle) is consistent with the deviation in shock slope from the Mach value.

The yawed flow past projectiles of revolution. Having plumbed the significance of Broderick's work on the cone, we pass to the present author's theory [*40*] of the yawed flow past projectiles of revolution. He uses an expansion in powers of the angle of yaw α, and takes the axis of the body, rather than the stream direction, as the axis of his cylindrical coordinates; but a few remarks are desirable on how his conclusions could be deduced from the smooth slender body theory above, in which the axis of coordinates is in the direction of the incident stream.

This approach simplifies the treatment of the equations of motion but complicates the boundary conditions. To a first approximation (only) the cross section of the projectile by a plane perpendicular to the stream is a circle of radius $R(x)$. Assuming this it is not hard to see that the distance normal to the stream between adjacent cross sections is given by

$$\frac{dn}{dx} = R'(x) + \alpha \cos \theta \tag{7-41}$$

provided that (r, θ) are taken temporarily *not* as cylindrical coordinates but as polar coordinates in the plane of the cross section with pole at the center of the circle, with the line $\theta = 0$ in the direction in which the projectile is yawed. But when we go to a second approximation, it is, strictly speaking, necessary to include perturbation terms due to the fact that the cross section is really an ellipse of axis ratio $\cos \alpha$. These are not actually important in practice, but can be calculated by considering how far the condition on $\partial\varphi_1/\partial n$ has *not* been satisfied on the ellipse (owing to being satisfied exactly on the circle), and then correcting

$\partial\varphi_2/\partial n$ by just this amount, which may be done, for simplicity, to within the accuracy of the second approximation, on the circle itself.

Because of Eq. 7-41 and 7-16 there is an additional term

$$-\alpha R^2(x)r^{-1}\cos\theta$$

in φ_1 due to yaw, and so an additional term $-\alpha R^2(x)\zeta^{-1}$ in $w(\zeta)$. This produces terms in the particular integral (7-25), additional to those in Broderick's solution, in α, α^2, and α^3, which include (among others) terms due to interaction of the yaw and nonyaw terms in φ_1. It should also be noticed that the *origin* of ζ varies from cross section to cross section (since it is always on the axis of the projectile, which is yawed to the x axis), and that therefore, in writing down the first and second terms of Eq. 7-25, terms due to the dependence of ζ on x (in such a way that $\partial\zeta/\partial x = -\alpha$) have to be included. Harmonics (with zero gradient at infinity) must then be added to make φ_2 satisfy the boundary condition (7-26) (with perhaps an additional term, see above, due to the ellipticity of the true cross section) on the circular boundary. This modifies the coefficients of α, α^2, and α^3. The term independent of r and θ is then modified, as discussed in the general theory. After this the term independent of α in φ_2 becomes exactly Broderick's term. Finally, the term in $r\cos\theta$ must be modified, by subtraction of a multiple of the harmonic $(r + r^{-1})\cos\theta$ representing incompressible flow past the circle, so that the coefficient $E(x)$ of $r\cos\theta$ takes the form (7-31) with $B(x) = -\alpha R^2(x)$.

Thus, whereas φ_1 contains terms *only* in 1 and α, φ_2 contains (besides Broderick's term in 1) terms in α, α^2, and α^3. These were all[7] obtained by the author [*40*], in his system of axes, though his method of presentation hardly made it clear that they form part of the same second approximation. He calculated the lift force on the part of the projectile given by $0 < x < l$ as

$$L = \rho_\infty q_\infty^2 S(l)\alpha\left\{1 + \frac{7}{6}(M_\infty^2 - 1)\alpha^2 + \frac{2M_\infty^2 - 1}{\pi}\left[S''(l)\ln\frac{(M_\infty^2 - 1)^{\frac{1}{2}}R(l)}{2l} + \int_0^l \frac{S''(l) - S''(y)}{l - y}\,dy\right] + \frac{M_\infty^2}{\pi}S''(l) + (M_\infty^2 - 2)R'^2(l)\right\} + 3(M_\infty^2 - 1)D_0\alpha \quad (7\text{-}42)$$

where D_0 is the drag at zero yaw on that part of the projectile (still ignoring the effect of the boundary layer), given by

$$D_0 = \frac{\rho_\infty q_\infty^2}{2\pi}\left[\int_0^l S''(x)dx\int_0^x S''(y)\ln\frac{(M_\infty^2 - 1)^{\frac{1}{2}}R(l)}{2(x - y)}\,dy - S'(l)\int_0^l S''(y)\ln\frac{(M_\infty^2 - 1)^{\frac{1}{2}}R(l)}{l - y}\,dy\right] \quad (7\text{-}43)$$

[7] **Except the term in α^2 independent of r and θ; this is not needed in calculating lifts and moments; in any case it is clear in principle, from the general account, how it could be obtained.**

Here the lift L signifies force normal to the stream, which is less by an amount $D_0\alpha$ than the force N normal to the axis of the projectile. To obtain the part of the *pressure distribution* which produces a force normal to the axis of the projectile, namely, the coefficient of $(-\cos\theta)$ when the pressure is Fourier analyzed, write x for l in $N = L + D_0\alpha$, differentiate with respect to x to obtain the normal force on the projectile between planes x and $x + dx$ normal to the axis, and divide by $\pi R(x)$. This may be used to determine the yawing moment [*40*].

It is seen from Eq. 7-42 that only small deviations of the lift curve from linearity can be attributable to the potential flow, since the proportional error in taking Eq. 7-42 as a multiple of α is about $\frac{7}{8}(M_\infty^2 - 1)\alpha^2$, which for large yaws like 10° is still only ten per cent for $M_\infty = 2$. We may therefore confine our attention to the predicted lift curve *slope*.[8]

A comparison with the exact adiabatic flow is possible for the case of a cone. Kopal [*37*] gives values of a coefficient K_N, which is, according to his equations for it (though his definition is rather loosely expressed),

$$K_N = \frac{1}{4R^2(l)\rho_\infty q_\infty^2}\left(\frac{dN}{d\alpha}\right)_{\alpha=0} \tag{7-44}$$

On the author's theory

$$K_N = \frac{1}{4}\pi\left[1 + M_\infty^2\delta^2 \ln\frac{(M_\infty^2 - 1)^{\frac{1}{2}}\delta}{2} + \left(\frac{3}{2}M_\infty^2 - 1\right)\delta^2\right] \tag{7-45}$$

where δ is the semiangle. In Fig. E,7e the plain lines indicate the exact values of K_N, the dashed lines denote his approximation, the first order approximation $K_N = \frac{1}{4}\pi$ is shown by a dotted line, and the dash-dotted line represents Tsien's prediction [*50*], based on a more exact solution of the linearized equation of motion. The author's theory is almost perfect for 5° semiangle, is good for $M_\infty \leqq 2$ with 10° semiangle, and is beginning to be a serious overestimate for 15° semiangle. On the other hand no advantage emerged from Tsien's taking a purely linear theory beyond the first approximation.[9]

For a body shape with a cylindrical portion at the rear, $R'(l)$ and $S''(l)$ are zero and Eq. 7-42 takes the simpler form

$$\left(\frac{dL}{d\alpha}\right)_{\alpha=0} = \rho_\infty q_\infty^2 S(l)\left[1 - \frac{2M_\infty^2 - 1}{\pi}\int_0^{l_1}\frac{S''(y)dy}{l - y}\right] + 3(M_\infty^2 - 1)D_0\alpha \tag{7-46}$$

where $x = l_1$ is the shoulder and the cylindrical portion is given by $l_1 < x < l$. Notice that *both* terms additional to the first approximation

[8] In fact there is very considerable departure from linearity for larger angles of attack, in the direction of higher lift. But this is due to boundary layer separation (see note 10).

[9] But for better approximations, in which the nonlinearity of the equation of motion for the cross flow is still not included, see Van Dyke [*54*].

$\rho_\infty q_\infty^2 S(l)$ for lift curve slope are positive. In the case of the integral this is because, though the integral of $S''(y)$ between the limits would be zero, the positive values of $S''(y)$ near the nose are weighted less by the factor $(l - y)^{-1}$ than the negative values farther back. This conclusion, that the lift curve slope always exceeds $\rho_\infty q_\infty^2$ times the base area, is in agreement

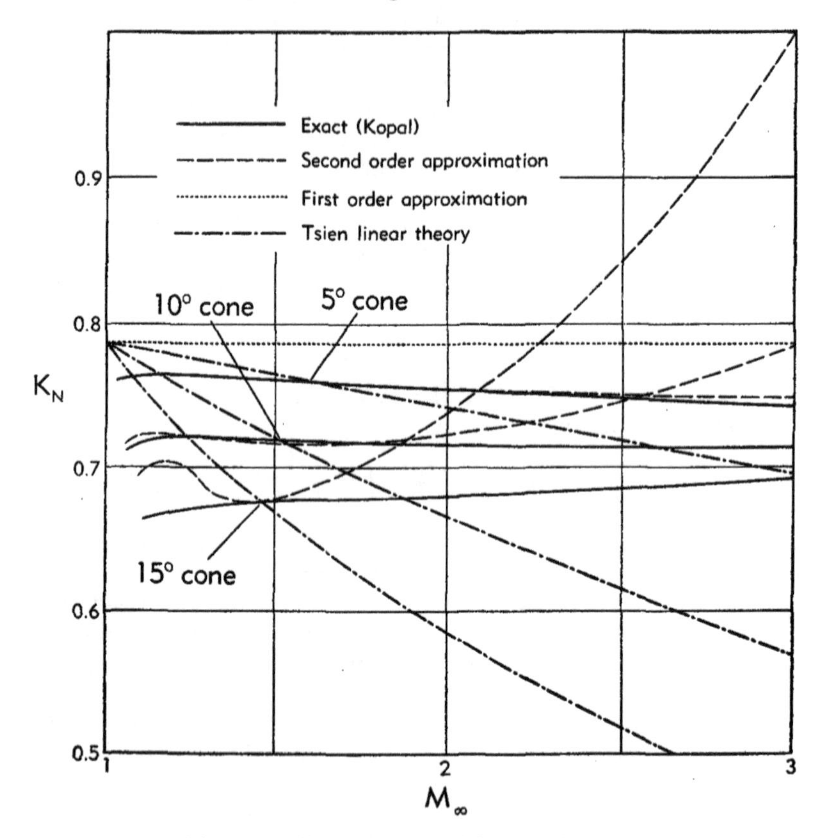

Fig. E,7e. Normal force coefficient K_N for a cone in slightly yawed flow at Mach number M_∞.

with experimentally determined lifts on projectiles, even though these contain an extra negative term due to the component normal to the stream of the skin friction and the base drag. It is also found, as the formula predicts, that the lift curve slope, divided by $\rho_\infty q_\infty^2$, increases with Mach number.

From a more quantitative point of view, take $D_0 = 0.8\rho_\infty q_\infty^2 S^2(l)/l_1^2$, (a typical value, about 25 per cent in excess of its theoretical minimum for given $S(l)$ and l_1), and estimate the integral in Eq. 7-46, by expanding $(l - y)^{-1}$ as $l^{-1} + yl^{-2}$, as $-S(l)/l^2$. This gives the approximation

$$\left(\frac{dL}{d\alpha}\right)_{\alpha=0} = \rho_\infty q_\infty^2 S(l)\left[1 + \frac{2M_\infty^2 - 1}{\pi}\frac{S(l)}{l^2} + 2.4(M_\infty^2 - 1)\frac{S(l)}{l_1^2}\right] \quad (7\text{-}47)$$

where $S(l)$ is the cross-sectional area of the cylindrical portion, whose length is $l - l_1$ out of a total length l. Clearly, the third term is more important than the second.

For a shell with 7.4-caliber radius ogival head, and total length 4.9 calibers, on which experiments have been made, $l = 9.8R(l)$ and $l_1 = R(l)\sqrt{4 \times 7.4 - 1}$. Eq. 7-47 then gives for the "cross wind force coefficient" commonly used in ballistics

$$f_L = \frac{1}{\rho_\infty q_\infty^2 R^2(l)}\left(\frac{dL}{d\alpha}\right)_{\alpha=0} = \pi\left\{1 + \frac{2M_\infty^2 - 1}{\pi}\frac{\pi R^2(l)}{[9.8R(l)]^2} + 2.4\,(M_\infty^2 - 1)\frac{\pi R^2(l)}{28.6R^2(l)}\right\} \quad (7\text{-}48)$$

$$= 3.17 + 0.89(M^2 - 1)$$

This (which neglects the negative component of skin friction and base drags) is 3.79 and 4.56 for $M_\infty = 1.3$ and 1.6 respectively, whereas the values measured by Cope and Thurston in wind tunnel tests (unpublished) were 3.70 and 3.90 respectively. (The error in the theory would be expected to increase noticeably with M_∞, even for such low values, since the nose semiangle is as high as 21°, which is comparable with the Mach angle 39° when $M_\infty = 1.6$.)

On the cylindrical portion of a projectile the part of the pressure distribution proportional to $\cos\theta$ is very small. For it is

$$-\frac{1}{\pi R(l)}\frac{\partial}{\partial l}\left(\frac{dN}{d\alpha}\right)_{\alpha=0} = -\frac{1}{\pi R(l)}\frac{\partial}{\partial l}\left[\left(\frac{dL}{d\alpha}\right)_{\alpha=0} + D_0\right]$$

$$= -\rho_\infty q_\infty^2 S(l)\frac{2M_\infty^2 - 1}{\pi^2 R(l)}\int_0^{l_1}\frac{S''(y)dy}{(l - y)^2} \quad (7.49)$$

by Eq. 7-46. This is at most a second order term, and would be about $\rho_\infty q_\infty^2(4M_\infty^2 - 2)[R(l)/l]^3$ near the base, or less than one per cent of $\rho_\infty q_\infty^2$ in typical cases. The only *first* order terms remaining in the pressure are

$$\rho_\infty q_\infty^2\left[\frac{1}{2\pi}\int_0^{l_1}\frac{S''(t)}{l - t}\,dt + \alpha^2\left(-\frac{1}{2} + \cos 2\theta\right)\right] \quad (7\text{-}50)$$

These represent a two-peaked variation about a negative mean. Near the nose, however, the term in α, representing a one-peaked variation (like $\cos\theta$), will normally exceed that in α^2 in importance, and the mean, represented by the axisymmetrical pressure distribution, will be positive. The transition from the one form to the others is well exemplified in the pressure plotting measurements of Cope and Thurston, which are shown in Fig. E,7f. The distance between peak and trough at the base is $0.13p_\infty$.

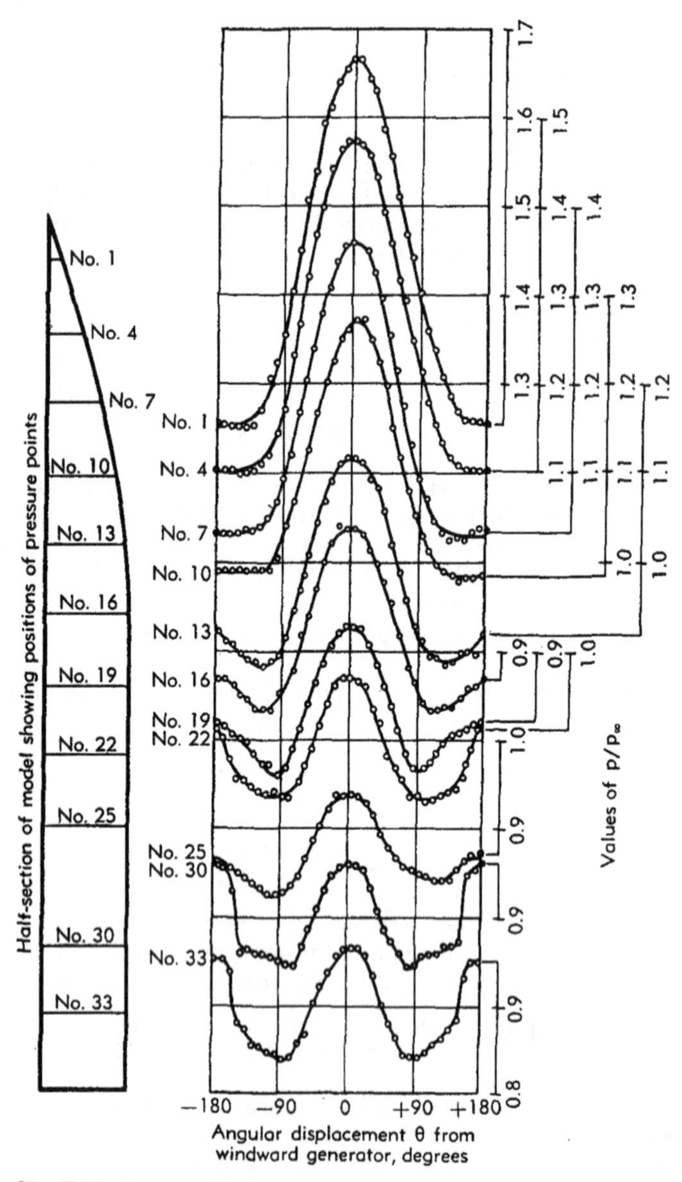

Fig. E,7f. Pressure distribution on a shell of six-caliber head radius in flow at Mach number 1.24, yawed to the shell by 10°.

which is precisely the value $2\rho_\infty q_\infty^2 \alpha^2$ predicted by Eq. 7-44. Note that the term in α^2 in Eq. 7-50 is that associated with the incompressible potential flow of a stream with velocity $q_\infty \alpha$ normal to the cylinder. It is amusing that, among the various boundary layer control devices which permit the incompressible potential flow normal to a cylinder to be achieved in practice, its superposition on a supersonic stream parallel to the axis of the cylinder is among the most effective![10] It should also be noticed that, if the shell is spinning, this cross flow produces a Magnus effect tending to yaw the shell about an axis at 90° to that of the existing yaw.

E,8. Supersonic Three-dimensional Wing Theory. The extension of the ideas of the foregoing articles to fully three-dimensional problems, that is to problems where no approximate dependence on only two space variables can be assumed, even in part of the flow pattern, has not yet been achieved. In particular, in three-dimensional wing theory, the surface pressure distribution has not been found to a better approximation than the linear theory gives, nor has much of the complete flow pattern been elucidated. It may be thought that the investigation must necessarily be too complicated to be worthwhile; but this is not so. For example if, by arduous calculations in special cases, one were to find that a value of C_p on linear theory signifies a true value considerably nearer to the value $C_p + (C_2/C_1^3)C_p^2$, as Eq. 3-21 shows is true in two dimensions, then this could be used to improve greatly our knowledge of aerodynamic forces. Again, in experiments on shocks produced by wedges in supersonic streams, it would be desirable to know whether any departures from two-dimensional shape in the wedge are fully reproduced in the shock, or whether they disappear far from the leading edge. Here we make no attempt to predict whether advances in these fields will become possible from the application of improvements in the physical approaches which are principally used in this section, or from formal power series expansions, combined with the use of the author's mathematical technique (for removing their divergences) near shocks, etc.

At present the only nonlinear studies of three-dimensional wing theory have been made in relation to the regions of "conical flow" (Busemann's "kegelige Strömungsfelder" *[46]*) which, for many wing shapes, consti-

[10] The reason, of course, is that the boundary layer is swept off at the rear before it has time to separate. The cross flow, which has velocity $U\alpha$, takes a time approximately $\frac{1}{3}R(l)/U\alpha$ to separate [*55*, p. 185], and therefore will not do so if αl is less than about $\frac{1}{3}R(l)$. However, for longer projectiles, or higher angles of attack, the ordinary gradual separation process for low speed flow past a circular cylinder occurs in the cross flow. The resulting suction on the leeward side near the base adds to the lift (of which the remainder comes from the nose) and decreases the destabilizing yawing moment. But there is some experimental evidence that unfavorable complications of this regime occur when $U\alpha$ reaches about half the speed of sound, so that (Art. 2) the shock stall for the cross flow is attained.

tute the front portion of the flow pattern. (But since in these regions the flow quantities are independent of distance from the apex, the problems are not "fully three-dimensional" in the sense of the last paragraph.) It has been seen in Sec. D that on linear theory, for many triangular and quadrilateral planforms with symmetrical sections, the whole contribution to the surface pressures due to angle of attack can be obtained by evaluating a conical field, or superposing two or three such fields. However, where nonlinear effects are considered, the division of the problem into the effect of thickness and of angle of attack is not permissible, nor is the superposition of solutions. Hence the nonlinear theory of conical fields can be applied only to those parts of a flow field which are fully conical, which means that those parts of the boundary which influence the field must be generated by straight lines through a single apex. Thus the influence of angle of attack on the flow past flat plane wings, with planforms of the triangular or quadrilateral shapes alluded to above, can be so treated. Also the effect of thickness can be taken into account, *but only over the front portion of the wing*, if similar sections at different spanwise stations, arranged so as to meet this condition, for example wedge sections with uniform wedge angle, are used in that portion.

The study of the subject was begun by the present author [*33*] who used his general technique [*31*] to determine the strength of the shock (of conical shape, nearly but not quite right circular) in an arbitrary conical field, given only the solution to the problem on linear theory. This would appear to be an investigation of only theoretical interest, but when F. K. Moore [*47*] went on to determine the surface pressures to the second approximation in a special case, he found that the solution was not unique unless a certain boundary condition, derived from the author's theory, was applied on the Mach cone from the apex. When previously a simpler boundary condition was applied (in Moore's Ph.D. thesis), the resulting incorrect second order solution for the surface pressures differed far more widely from the linear theory than does the true solution. Moore's theory has since been amplified by Tan [*48*], who computes the surface pressures in a wider range of cases.

These three references constitute the whole existing literature on the subject of the present article, although further work is in progress, particularly in France on the use of an electric analogy to calculate the second approximate flow field, and in England on the problem of improving the linear theory for delta wings with rounded leading edge (inside the Mach cone from the apex) at angle of attack, in order to deduce a finite pressure distribution around the leading edge. This problem is similar to one encountered in subsonic flow (Art. 2), since the flow normal to the leading edge is subsonic, and it can perhaps be treated by similar methods.

The cone field theory will here be discussed only for cases when the wing lies entirely within the Mach cone from the apex. The reader is referred to [*33*] for a discussion of the shock strength in the general case, and in particular of the interesting shock behavior, near the tip of a rectangular wing at angle of attack, where the Prandtl-Meyer fan of expansion waves on the upper surface interacts with the tip shock, pulling it in toward the axis in a most peculiar corrugation. (See also [*57,58*].)

With this restriction, the author's conclusions will here be deduced by a simpler physical argument, based on the principle enunciated near the end of Art. 7. (Of course this is the more convincing for being supported by his original mathematical argument.) They will then be used in deriving the Moore-Tan theory of the surface pressures.

It is the general conclusion of this section that an adiabatic flow field can, even behind shocks, be evaluated to good approximation by solving the equations of potential motion. But it has been seen that the complete solution of these equations has in general singular regions, at each point of which it predicts three different values for each flow quantity. These regions must in practice be traversed by a shock, ahead of which the flow quantities take one of the three values, and behind which they take one of the others. For the two-dimensional flow past a sharp-nosed airfoil, the singular region associated with the front shock is of wedge shape, being bounded by a Mach line associated with the flow deflection at the nose on the upper surface, and by another associated with the undisturbed stream. For axisymmetric flow past a cone the region is bounded by the undisturbed Mach cone and by a cone of larger angle, which may be called the limit cone (being the upstream limit of disturbances in the potential flow) and which is an envelope of Mach lines. Such an envelope (not conical) also exists for any supersonic flow past slender projectiles, and may be obtained in the usual way by eliminating y from Eq. 6-6 and from the same equation differentiated with respect to y.

Similarly, it may be supposed that the exact solution, for potential flow, to our conical field problem has such a singular region, bounded behind by the undisturbed Mach cone from the apex, and ahead by a limit cone (not, of course, necessarily right circular), and that the shock cone must be placed in this region. The flow (and hence finally the shock position) is then deduced by applying word for word the principle stated in italics on p. 470, and expressing the condition that, since the limit cone is the forward boundary of the potential flow field, the component of velocity normal to it must equal the local speed of sound (for the limit cone, if not a stationary wave front itself, is at least an envelope of such wave fronts).

The first step is to write down the equations for the first and second approximate solutions. The first approximation φ_1 to the disturbance potential satisfies the linearized equation, while the second approximation, by Eq. 2-14, satisfies

$$\frac{\partial^2\varphi_2}{\partial r^2}+\frac{1}{r}\frac{\partial\varphi_2}{\partial r}+\frac{1}{r^2}\frac{\partial^2\varphi_2}{\partial\theta^2}-(M_\infty^2-1)\frac{\partial^2\varphi_2}{\partial x^2}$$
$$=M_\infty^2\frac{\partial}{\partial x}\left[\left(1+\frac{\gamma-1}{2}M_\infty^2\right)\left(\frac{\partial\varphi_1}{\partial x}\right)^2+\left(\frac{\partial\varphi_1}{\partial r}\right)^2+\frac{1}{r^2}\left(\frac{\partial\varphi_1}{\partial\theta}\right)^2\right] \quad (8\text{-}1)$$

in cylindrical coordinates. The solution of the linearized equation for φ_1 is most easily accomplished in terms of a variable

$$s=\cosh^{-1}\left(\frac{x}{(M_\infty^2-1)^{\frac{1}{2}}r}\right) \quad (8\text{-}2)$$

for then, in a cone-field problem, in which the disturbance velocities $u=\partial\varphi/\partial x$, $v=\partial\varphi/\partial y$, and $w=\partial\varphi/\partial z$ can depend only on nondimensional variables like s and θ, they are actually harmonic functions of s, θ, i.e. solutions of $\partial^2u/\partial s^2+\partial^2u/\partial\theta^2=0$. The determination of these to satisfy the boundary conditions and vanish on the undisturbed Mach cone $s=0$ is (D, 13) a fairly easy matter. The behavior of these velocities near the cone is given by equations like

$$\frac{\partial\varphi_1}{\partial x}\sim -A(\theta)\frac{s}{\sqrt{2}}\sim -A(\theta)\sqrt{\frac{x}{(M_\infty^2-1)^{\frac{1}{2}}r}-1} \quad (8\text{-}3)$$

where the function $A(\theta)$ is positive if the pressure increases (on linear theory) as the fluid passes the undisturbed Mach cone, and negative if it decreases. The forms for φ_1 and $\partial\varphi_1/\partial r$ near the Mach cone follow at once, whence the right-hand side of Eq. 8-1 becomes, in the limit as $x\to(M_\infty^2-1)^{\frac{1}{2}}r$,

$$M_\infty^2\frac{\partial}{\partial x}\left\{\left(1+\frac{\gamma-1}{2}M_\infty^2\right)A^2(\theta)\left[\frac{x}{(M_\infty^2-1)^{\frac{1}{2}}r}-1\right]\right.$$
$$\left.+(M_\infty^2-1)A^2(\theta)\left[\frac{x}{(M_\infty^2-1)^{\frac{1}{2}}r}-1\right]\right\}=\frac{(\gamma+1)M_\infty^4}{2(M_\infty^2-1)^{\frac{1}{2}}}\frac{A^2(\theta)}{r} \quad (8\text{-}4)$$

Now if φ_2 is taken continuous, and hence zero, on the undisturbed Mach cone $x=(M_\infty^2-1)^{\frac{1}{2}}r$, then both the third term, and the sum of the first and fourth terms, of the left-hand side of Eq. 8-1 vanish on the cone. Hence by Eq. 8-4

$$\frac{\partial\varphi_2}{\partial r}=\frac{(\gamma+1)M_\infty^4}{2(M_\infty^2-1)^{\frac{1}{2}}}A^2(\theta),\qquad \frac{\partial\varphi_2}{\partial\theta}=0 \quad (8\text{-}5)$$

on the cone, and the other nonvanishing component of velocity $\partial\varphi_2/\partial x$ follows in the form

$$\frac{\partial\varphi_2}{\partial x}=-\frac{r}{x}\frac{\partial\varphi_2}{\partial r}=-\frac{(\gamma+1)M_\infty^4}{2(M_\infty^2-1)}A^2(\theta) \quad (8\text{-}6)$$

from the fact that the field is conical.

The basic principle of the present method now tells us that Eq. 8-5 and 8-6 represent, not the velocity on the undisturbed Mach cone, but the velocity on the limit cone. The equation of the limit cone must take the general form

$$x = (M_\infty^2 - 1)^{\frac{1}{2}} r[1 - G(\theta)] \tag{8-7}$$

where $G(\theta)$ is small and positive. Expressing the condition that the component of velocity normal to this surface (calculated from the disturbance velocities (8-5) and (8-6)) is equal to the local speed of sound (calculated from Bernoulli's equation), and neglecting the squares of $G(\theta)$, $G'(\theta)$, φ_x, and φ_r, we obtain

$$G(\theta) = \frac{(\gamma + 1)^2 M_\infty^8}{4(M_\infty^2 - 1)^2} A^2(\theta) \tag{8-8}$$

Eq. 8-7 and 8-8 determine the limit cone as differing from the undisturbed Mach cone by an angle of the order of the square of the disturbances in the main part of the flow.

The approximation to $\partial\varphi/\partial x$ near the undisturbed Mach cone is given as the sum of Eq. 8-3 and 8-6, two terms which are of the same order of magnitude when the quantity under the square root in Eq. 8-3 is of the order of Eq. 8-8. But according to the basic principle, this expression would be correct only if the distance x were given a new origin, so that the curve $x - (M_\infty^2 - 1)^{\frac{1}{2}} r = 0$ became the limit cone. Thus, if no such alteration in the meaning of x is made, the true form of $\partial\varphi/\partial x$ in this neighborhood satisfies

$$\frac{\partial\varphi}{\partial x} \cong -A(\theta) \sqrt{\frac{x}{(M_\infty^2 - 1)^{\frac{1}{2}} r} - 1 + G(\theta)} - \frac{(\gamma + 1) M_\infty^4}{2(M_\infty^2 - 1)} A^2(\theta) \tag{8-9}$$

To the same order $\partial\varphi/\partial r = -(M_\infty^2 - 1)^{\frac{1}{2}} \partial\varphi/\partial x$.

The next term $\partial\varphi_3/\partial x$ would of course be infinite on $r = 1$ in successive approximation theory, but it is essential to the principle used that this is no longer so when the change in the equations due to shifting the origin of x (which begins to affect the φ_n for $n = 3$) is taken into account. The author [*33*] proves that this is in fact so. Hence Eq. 8-9 represents the terms of highest order in $\partial\varphi/\partial x$ near $x = (M_\infty^2 - 1)^{\frac{1}{2}} r$.

The shock must also be a cone, with equation say

$$x = (M_\infty^2 - 1)^{\frac{1}{2}} r[1 - H(\theta)] \tag{8-10}$$

The longitudinal disturbance velocity (8-9) behind it is related to the angle which it makes with the undisturbed stream. Expressing this relation, with terms of smaller order than $A^2(\theta)$ neglected, we have

$$\frac{4(M_\infty^2 - 1)}{(\gamma + 1) M_\infty^4} H(\theta) = A(\theta) \sqrt{G(\theta) - H(\theta)} + \frac{(\gamma + 1) M_\infty^4}{2(M_\infty^2 - 1)} A^2(\theta) \tag{8-11}$$

By suitable squaring, Eq. 8-11 can evidently be reduced to a quadratic equation in $H(\theta)$. But of the roots of this equation, only one is a solution of Eq. 8-11; the other satisfies the same equation with the sign changed in front of the square root. Actually, one of the said two roots is $H(\theta) = 0$. By Eq. 8-8 this is a solution of Eq. 8-11 when $A(\theta) < 0$. The other root, appropriate when $A(\theta) > 0$, is

$$H(\theta) = \frac{3}{4} G(\theta) = \frac{3}{16} \frac{(\gamma + 1) M_\infty^8}{(M_\infty^2 - 1)^2} A^2(\theta) \tag{8-12}$$

In Eq. 8-10 and 8-12 we have expressed the shock position, to the second order in the disturbances, in terms of the behavior (see Eq. 8-3) of the linearized solution to the cone field problem near the undisturbed Mach cone. But this expression has been shown to be the correct one only when $A(\theta) > 0$, i.e. when on the linear theory the pressure increases as the fluid passes the undisturbed Mach cone. When this is not so $H(\theta)$ is zero and the shock strength is at most of the order of the cube of the disturbances (the author [*33*] gives reasons for supposing it not to be exactly zero). The correspondence of this mathematical point with the physical fact that shock discontinuities are necessarily compressive is most interesting.

The pressure gradient along a streamline just behind the shock is easily calculated from Eq. 8-9 in each of the two cases. When $A(\theta) > 0$, so that Eq. 8-12 holds, it is given by

$$x \frac{\partial C_p}{\partial x} = \frac{2(M_\infty^2 - 1)}{(\gamma + 1) M_\infty^4} \tag{8-13}$$

which is not small, though pressure gradients of this magnitude act only over a small distance behind the shock. This thin region of rapid compression (not of course large enough to bring viscosity into play) "reinforces" the shock (itself a region of almost infinitely rapid compression). But when $A(\theta) < 0$, and so the shock is much weaker, the pressure gradient just behind it is given by

$$x \frac{\partial C_p}{\partial x} = - \frac{(M_\infty^2 - 1)}{(\gamma + 1) M_\infty^4} \tag{8-14}$$

and falls to half this value (for example) when $x/(M - 1_\infty^2)^{\frac{1}{2}} r$ is about $1 - 3G(\theta)$. The pressure drop in this region would probably be the most noticeable phenomenon on an interferometer record, as it exceeds the pressure rise at the shock. It is the analogue of the Prandtl-Meyer expansion in two-dimensional flow, in which also the pressure gradient is not small even though the total expansion be small.

When $A(\theta) > 0$ the strength of the shock is given, to the approximation considered, by

$$\frac{\Delta p}{p} = \frac{3}{4} \gamma(\gamma + 1) \frac{M_\infty^6}{M_\infty^2 - 1} A^2(\theta) \tag{8-15}$$

The author [33] gave two examples of Eq. 8-15. In one the obstacle was slender in all directions perpendicular to the stream. In this case $A(\theta) = (\Omega/\pi)\sqrt{2}$, in terms of the solid angle Ω at the apex. The resulting value for the strength agrees with that which follows from Whitham's later theory of Art. 6. In the second example the obstacle was a flat delta wing at angle of attack α. In the unyawed case, when the leading edges make an angle $\frac{1}{2}\pi - \tau$ with the incident stream, the shock strength below the wing is a multiple of α^2 varying with M_∞ like $M_\infty^6/(M_\infty^2 - 1)^2$ and with θ like $\sin^2\theta[1 - (M_\infty^2 - 1)\cot^2\tau\cos^2\theta]^{-3}$; above the wing it is of smaller order than α^2. There is one maximum of shock strength, in the middle, when $\tau > \tan^{-1}\sqrt{3(M_\infty^2 - 1)}$. For example when $M_\infty = \sqrt{2}$ and $\tau = 67.5°$ this maximum shock strength is only 0.026 for $\alpha = 10°$. But when $\tau < \tan^{-1}\sqrt{3(M_\infty^2 - 1)}$ the strength rises to a maximum on two symmetrically placed generators with $\cos^2\theta = \frac{1}{2}[3 - (M_\infty^2 - 1)^{-1}\tan^2\tau]$, while the central generator is a local minimum of strength. For example when $M_\infty = \sqrt{2}$, $\tau = 55°$, and $\alpha = 10°$ the maximum strength is 0.07; and this continues to increase as the leading edge approaches the undisturbed Mach cone.

We pass now to the calculation of the full second approximation to the flow field, especially on the body surface. The boundary condition appropriate on the Mach cone is shown by the above theory to be the very natural one obtained by using Eq. 8-1 for the second approximation also in the limit as $x \to (M_\infty^2 - 1)^{\frac{1}{2}}r$. The velocity vector on the Mach cone is in fact given by Eq. 8-5 and 8-6. In terms of the velocities $u_2 = \partial\varphi_2/\partial x$, $v_2 = \partial\varphi_2/\partial y$, $w_2 = \partial\varphi_2/\partial z$ which are found most convenient for solving the equations, this boundary condition becomes (using the definition (8-3) of $A(\theta)$)

$$u_2 = -\frac{(\gamma+1)M_\infty^4}{M_\infty^2 - 1}\left(\frac{\partial u_1}{\partial s}\right)_{s=0}^2, \qquad v_2 = \frac{(\gamma+1)M_\infty^4}{(M_\infty^2 - 1)^{\frac{1}{2}}}\left(\frac{\partial u_1}{\partial s}\right)_{s=0}^2 \cos\theta,$$

$$w_2 = \frac{(\gamma+1)M_\infty^4}{(M_\infty^2 - 1)^{\frac{1}{2}}}\left(\frac{\partial u_1}{\partial s}\right)_{s=0}^2 \sin\theta \tag{8-16}$$

But since the homogeneous part of Eq. 8-1, for the second approximation to the potential, is the same as the equation for the first approximation, the same is true of the equations for the velocities. But in the variables s (Eq. 8-2) and θ, Laplace's equation is satisfied by u_1, v_1, and w_1, and therefore equations of Poisson's type are satisfied by u_2, v_2, and w_2. Moore [47] worked out these equations, and also the relations which must subsist between u_2, v_2, and w_2 in the s, θ plane due to irrotationality. By working with complex variables he then obtains a particular integral for u_2 (rather as Eq. 7-22 was obtained). To this he has merely to add a solution of the ordinary linear cone-field equations, so that the whole satisfies the boundary conditions.

When the wing lies approximately in the plane $z = 0$, the boundary conditions thereon can be approximated (by using Taylor's theorem) in the form of a statement of the values of $\partial\varphi_1/\partial z = w_1$ and $\partial\varphi_2/\partial z = w_2$ on the plane $z = 0$. Using the irrotationality conditions we may then infer the boundary conditions on u_1 and u_2 on $z = 0$, and the kind of singularities which they may be permitted to satisfy at the leading edge. The second approximation may then be obtained uniquely using finally that Eq. 8-16 holds on $s = 0$. (The detailed process of calculation is not here discussed, as it is still in such an early stage that great improvements are certain to occur later on.) From u_2 and the first approximation the pressure coefficient follows in the form

$$C_p = -2u_1 - [2u_2 - (M_\infty^2 - 1)u_1^2 + v_1^2 + w_1^2] \tag{8-17}$$

Moore [47] and Tan [48] both considered, as an example of the method, the unyawed flow past a delta wing, whose airfoil section is a wedge of semiangle ϵ, at zero angle of attack. The cross section of the wing by a plane normal to the stream is thus a rhombus, whose center is the foot of the perpendicular from the apex to the said plane. If the planform makes an angle 2λ at the apex, the diagonals of this rhombus are in the ratio $\tan \epsilon : \tan \lambda$.

Moore considered only a single case, but Tan calculated the pressure distribution for a series of values for M_∞, λ, and ϵ. He took $M_\infty = 1.2$, $\sqrt{2}$ $(= 1.414)$, and 2, and used four different values for the parameter $(M_\infty^2 - 1)^{\frac{1}{2}} \tan \lambda$, which is the ratio of wing span to diameter of Mach cone on any plane normal to the stream. Since the wing is inside the Mach cone this parameter must be less than 1, and the four values used for it were 0.3651, 0.5505, 0.8, and 0.9535.

Tan's graphs of pressure distribution show quite clearly that the relation of the second approximation to the first remains closely similar for different M_∞ if the parameter $(M_\infty^2 - 1)^{\frac{1}{2}} \tan \lambda$ is kept constant. To be precise, at points whose distances from the axis of the Mach cone are a given fraction of its local radius, the ratio of the two approximations to the pressure coefficient changes little with Mach number. To illustrate the general behavior we may therefore confine ourselves to the single value $\sqrt{2}$ for M_∞.

For this value of M_∞, for $\epsilon = 3°$, and for the four values of

$$(M_\infty^2 - 1)^{\frac{1}{2}} \tan \lambda$$

listed above (which for $M_\infty = \sqrt{2}$ are simply $\tan \lambda$ and correspond to $\lambda = 20°4'$, $28°50'$, $38°39'$, and $43°38'$) the first and second approximations to the distribution of pressure coefficient C_p along the surface are illustrated in Fig. E,8. In this figure we work in a plane normal to the incident stream, which cuts the Mach cone in a circle whose radius is taken as the unit of length. Then the plane cuts the wing in four alternative sections as illustrated (confining ourselves by symmetry to the right-hand half of each) at the bottom of the figure. The two approximations to C_p are graphed for each wing until just before the tip. At the tip C_p takes in reality a fairly large value. This may be estimated, by assuming that

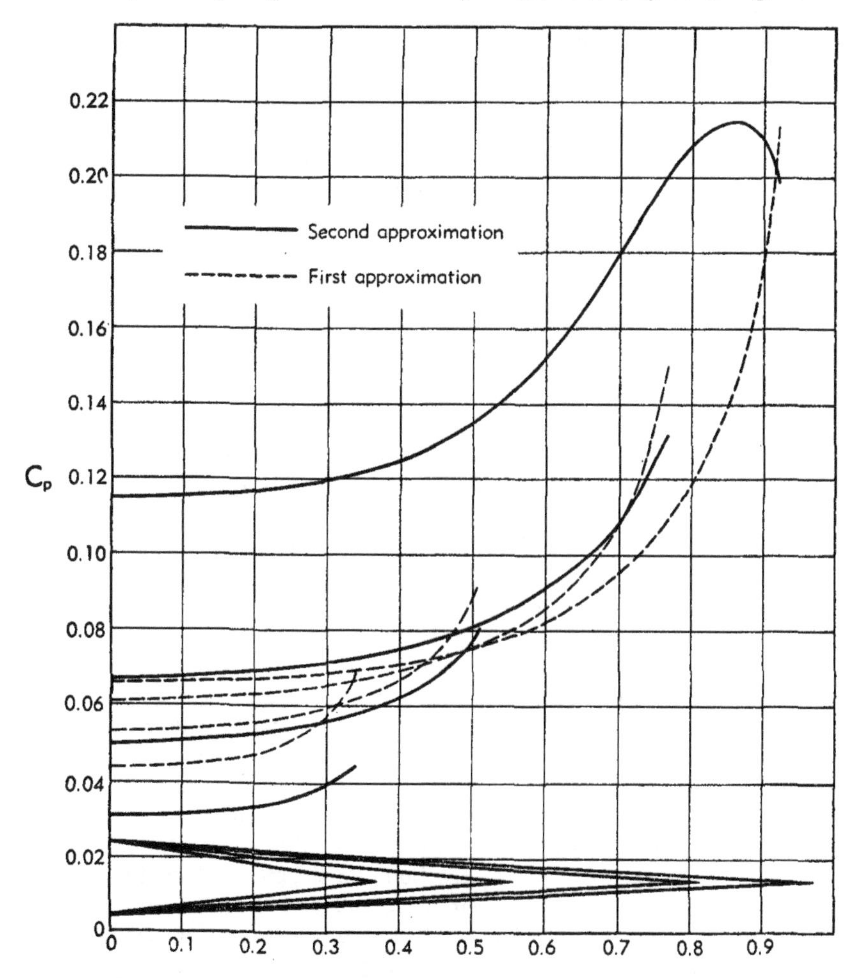

Fig. E,8. Pressure distribution on four delta wings, each with uniform wedge airfoil section of semiangle 3°, at Mach number $\sqrt{2}$.

the component of fluid velocity normal to the leading edge is reduced to zero there (by analogy with the subsonic flow past a wedge), while the component along the edge is as in the incident stream, as 0.125, 0.26, 0.47, and 0.60 in the four cases respectively. But, of course, no method of successive approximation could be expected to give the correct value here; in fact both approximations give integrable infinities for C_p at the tip. (There is no need to give the corresponding curves for other values of ϵ, the semiangle of the wedge airfoil section, since the first approximation to C_p is simply proportional to ϵ, and the difference between the two approximations is proportional to ϵ^2.)

We see from the figure that for $\epsilon = 3°$ the accuracy of the first approximation to the pressure distribution (as inferred from its difference from the second approximation) would be tolerable for practical purposes only in an intermediate range of $(M_\infty^2 - 1)^{\frac{1}{2}} \tan \lambda$, including 0.5505 and 0.8 but not 0.3651 and 0.9535. It is not in the least surprising that the error increases as $(M_\infty^2 - 1)^{\frac{1}{2}} \tan \lambda$ approaches 1, i.e. as the leading edge approaches the Mach cone (so that the flow normal to it is transonic) and this deficiency is far from remediable in the present state of knowledge. It is much odder, however, that the error should increase as λ decreases, so that when the planform has apex angle $2\lambda = 40°8'$ the error is no longer tolerable. It might be thought that this is due to the body becoming slender in all directions normal to the stream, as in Ward's theory (Art. 4), so that the terms in v_1^2 and w_1^2 in the expression (8-17) for C_p need to be taken into account even on a first approximation; but in fact λ is still much greater than ϵ, the terms in the squares of velocities in Eq. 8-17 are unimportant, and the difference u_2 in the axial perturbation velocities is just as considerable as the difference in the pressures.

This being so, it is very hard to believe that the linear theory should be in error as much as Fig. E,8 suggests, considering that the wing surface makes everywhere an angle less than 3° with the stream. It is probable, in fact, that the second approximation may in this case be farther from the truth than the first, since in its deduction from the first the singularity at the leading edges may play a dominant part, accentuated by the proximity of the leading edges for smaller λ and by the method of Taylor expansion to obtain boundary conditions on the plane $z = 0$. Thus we may, as indicated so often in this section, have run into difficulties by too readily using a formal expansion in series instead of proceeding from one approximation to another by physical reasoning.

It may nevertheless be the case that the theory is adequate in the intermediate range of $(M_\infty^2 - 1)^{\frac{1}{2}} \tan \lambda$, and if so it is very interesting that, as Tan remarks (interpolating between the curves for 0.5505 and for 0.8 in Fig. E,8), the first and second approximations to C_p are almost indistinguishable for $(M_\infty^2 - 1)^{\frac{1}{2}} \tan \lambda = 0.65$. This property is reproduced for $M_\infty = 1.2$ and for $M_\infty = 2$, and so may be supposed independent of Mach number. Whether it is also independent of airfoil section, so that the linear theory is always specially good for delta wings spanning two thirds of the Mach cone from the apex, is of course not known.

In future calculations on delta wings inside the Mach cone, airfoils with rounded leading edges should be used, since these are in practice essential to prevent stalling at low angles of attack. An example would be the right elliptical cone, for which the linear theory was calculated by Squire [*49*] and by Laporte and Bartels [*51*]. Boundary conditions should preferably be satisfied on the surface itself as in the Hantzsche-Wendt method, or if they are satisfied on the plane of symmetry a special technique should be applied to treat conditions near the leading edge (Art. 2).

There is little in Fig. E,8 to support the suggestion, made at the beginning of this article, that the two-dimensional result, "a given value of C_p on linear theory signifies a true value nearer to $C_p + (C_2/C_1^2)C_p^2$," might be true in some degree of generality. It is correct only for values of $(M_\infty^2 - 1)^{\frac{1}{2}} \tan \lambda$ around 0.8. However, although it is shown to be false, for a delta wing with leading edges markedly "subsonic," in the front portion where the field is conical, it might be expected to apply with greater accuracy to the pressure *changes* as the flow expands (more nearly two-dimensionally) around the rear portion of the wing (compare the expansion round the corner in Fig. E,7b).

If matters such as this are to be decided, the extension of the theories described in Sec. E to "fully three-dimensional" problems is necessary.

E,9. Cited References.

1. Janzen, O. Beitrag zu einer Theorie der stationären Strömung kompressibler Flüssigkeiten. *Phys. Zeits. 14*, 639 (1913).
2. Rayleigh, Lord. On the flow of compressible fluid past an obstacle. *Phil. Mag. 32*, 1 (1916).
3. Eser, F. Zur Strömung kompressibler Flüssigkeiten um feste Körper mit Unterschallgeschwindigkeit. *Luftfahrtforschung 20*, 220 (1943).
4. Imai, I. On the flow of a compressible fluid past a circular cylinder. *Proc. Phys. Math. Soc. Japan 20*, 636 (1938).
5. Tamada, K. On the flow of a compressible fluid past a sphere. *Proc. Phys. Math. Soc. Japan 21*, 743 (1939).
6. Kaplan, C. Compressible flow about symmetrical Joukowski profiles, A theoretical study of the moment on a body in a compressible fluid, *and* On the use of residue theory for treating the subsonic flow of a compressible fluid. *NACA Reports 621*, 1938; *671*, 1939; and *728*, 1942.
7. Goldstein, S., and Lighthill, M. J. Two-dimensional compressible flow past a solid body in unlimited fluid or symmetrically placed in a channel. *Phil. Mag. 35*, 549 (1944).
8. Tomotika, S., and Umemoto, H. On the subsonic flow of a compressible fluid past a symmetrical Joukowski aerofoil. *Rep. Aero. Res. Inst. Tokyo 16*, 35 (1941).
9. Taylor, G. I., and Sharman, C. F. A mechanical method for solving problems of flow in compressible fluids. *Proc. Roy. Soc. A121*, 194 (1928).
10. Görtler, H. Gasströmungen mit Übergang von Unterschall- zu Überschallgeschwindigkeiten. *ZAMM 20*, 254 (1940).
11. Hantzsche, W., and Wendt, H. Der Kompressibilitätseinfluss für dünne wenig gekrümmte Profile bei Unterschallgeschwindigkeit. *ZAMM 22*, 72 (1942).
12. Kaplan, C. Effect of compressibility at high subsonic velocities on the lifting force acting on an elliptic cylinder. *NACA Technical Note 1118*. 1946.
13. Glauert, H. *Elements of Aerofoil and Airscrew Theory*. Cambridge Univ. Press, 1926.
14. Theodorsen, T. Theory of wing sections of arbitrary shape, *NACA Report 411*. 1931.
15. Hantzsche, W. Die Prandtl-Glauertsche Näherung als Grundlage für ein Iterations-

verfahren zur Berechnung kompressibler Unterschallströmungen. *ZAMM 23*, 185 (1943).
16. Tsien, H. S. Two-dimensional subsonic flow of compressible fluids. *J. Aero. Sci. 6*, 399 (1939).
17. Prandtl, L. Neue Untersuchungen über die strömende Bewegung der Gase und Dämpfe. *Phys. Zeits. 8*, 23 (1907).
18. Ackeret, J. Luftkräfte auf Flügel die mit grösserer als Schallgeschwindigkeit bewegt werden. *Z. Flugtech. u. Motorluftschiffahrt 16*, 72 (1925).
19. Friedrichs, K. O. Formation and decay of shock waves. *Comm. Pure and Appl. Math. 1*, 211 (1948).
20. Busemann, A. Aerodynamischer Auftrieb bei Überschallgeschwindigkeit. *Luftfahrtforschung 12*, 210 (1935).
21. Epstein, P. S. On the air resistance of projectiles. *Proc. Nat. Acad. Sci. 17*, 532 (1931).
22. Donov, A. A plane wing with sharp edges in a supersonic stream. *Bull. Acad. Sci. URSS. Sér. Math. 1939*, 603 (1939).
23. Thomas, T. Y. The determination of pressure on curved bodies behind shocks. *Comm. Pure and Appl. Math 3*, 103 (1950).
24. Chandrasekhar, S. On the decay of plane shock waves. *Ballistics Research Lab. Report 423*. 1943.
25. Lighthill, M. J. The conditions behind the trailing edge of the supersonic aerofoil. *ARC Reports and Memoranda 1930*. 1944.
26. Martin, M. H. A problem in the propagation of shock. *Quart. Appl. Math. 4*, 330 (1947).
27. Meyer, R. E. The method of characteristics for problems of compressible flow involving two independent variables. *Quart. J. Mech. and Appl. Math. 1*, 196–219 (1948).
28. Lighthill, M. J. The flow behind a stationary shock. *Phil. Mag. 40*, 214 (1949).
29. Kahane, A., and Lees, L. The flow at the rear of a two-dimensional supersonic airfoil. *J. Aero. Sci. 15*, 167–170 (1948).
30. Lighthill, M. J. The energy distribution behind decaying shocks. I. Plane waves. *Phil. Mag. 41*, 1101 (1950).
31. Lighthill, M. J. A technique for rendering approximate solutions to physical problems uniformly valid. *Phil. Mag. 40*, 1179 (1949).
32. Whitham, G. B. The behaviour of supersonic flow past a body of revolution, far from the axis. *Proc. Roy. Soc. A201*, 89 (1950).
33. Lighthill, M. J. The shock strength in supersonic "conical fields." *Phil. Mag. 40*, 1202 (1949).
34. Broderick, J. B. Supersonic flow round pointed bodies of revolution, *and* Supersonic flow past a semi-infinite cone, *Quart. J. Mech. and Appl. Math. 2*, 98–120; 121–128 (1949).
35. Lighthill, M. J. Supersonic flow past slender bodies of revolution the slope of whose meridian section is discontinuous. *Quart. J. Mech. and Appl. Math. 1*, 90 (1948).
36. Lighthill, M. J. The position of the shock-wave in certain aerodynamic problems. *Quart. J. Mech. and Appl. Math. 1*, 309 (1948).
37. Kopal, Z. *Tables of Supersonic Flow Around Cones*, and *Tables of Supersonic Flow Around Yawing Cones*, Mass. Inst. of Tech., 1947.
38. Taylor, G. I., and Maccoll, J. W. The air pressure on a cone moving at high speeds. *Proc. Roy. Soc. A139*, 278 (1933).
39. DuMond, J. W. M., Cohen, E. R., Panofsky, W. K. H., and Deeds, E. A determination of the wave forms and laws of propagation and dissipation of ballistic shock waves. *J. Acoust. Soc. Amer. 18*, 97 (1946).
40. Lighthill, M. J. Supersonic flow past slender pointed bodies of revolution at yaw. *Quart. J. Math. and Appl. Math. 1*, 76 (1948).
41. von Kármán, Th., and Moore, N. B. Resistance of slender bodies moving with supersonic velocities, with special reference to projectiles. *Trans. ASME 54*, 303–310 (1932).
42. Lighthill, M. J. Supersonic flow past bodies of revolution. *ARC Reports and Memoranda 2003*. 1945.

43. Ward, G. N. Supersonic flow past slender pointed bodies. *Quart. J. Mech. and Appl. Math. 2*, 75 (1949).
44. Watson, G. N. *Theory of Bessel Functions*, 2nd ed. Cambridge Univ. Press, 1944.
45. Ward, G. N. The approximate external and internal flow past a quasi-cylindrical tube moving at supersonic speeds. *Quart. J. Mech. and Appl. Math. 1*, 225 (1948).
46. Busemann, A. Infinitesimale kegelige Überschallströmung. *Luftfahrtforschung 20*, 105 (1943).
47. Moore, F. K. Second approximation to supersonic conical flows. *J. Aero. Sci. 17*, 328 (1950).
48. Tan, H. S. *Second Approximation to Conical Flows*. Cornell Univ. Grad. School of Aero. Engrg., Ithaca, 1950. Also *WADC Technical Report 52-277*, Wright Air Dev. Center, Dayton, O., 1952.
49. Squire, H. B. Theory of the flow over a particular wing in a supersonic stream. *Roy. Air. Est. Rept. 2184*. Farnborough, 1947.
50. Tsien, H. S. Supersonic flow over an inclined body of revolution. *J. Aero. Sci. 5*, 480 (1938).
51. Laporte, O., and Bartels, R. C. F. An investigation of the exact solutions of the linearized equations for the flow past conical bodies. *Bumblebee Series Report 75*. University of Michigan, Ann Arbor, 1948.
52. Kaplan, C. The flow of a compressible fluid past a curved surface. *NACA Report 768*. 1943.
53. Kaplan, C. On the particular integrals of the Prandtl-Busemann iteration equations for the flow of a compressible fluid. *NACA Technical Note 2159*. 1950.
54. Van Dyke, M. First- and second-order theory of supersonic flow past bodies of revolution. *J. Aero. Sci. 18*, 161 (1951).
55. Goldstein, S. *Modern Developments in Fluid Dynamics*. Oxford Univ. Press, 1938.
56. Lighthill, M. J. A new approach to thin aerofoil theory. *Aero. Quart. 3*, 193 (1951).
57. Legras, M. J. Application de la méthode de Lighthill à un écoulement plan supersonique. *C.R. Acad. Sci. Paris 233*, 1005 (1951).
58. Legras, M. J. Écoulement conique au voisinage d'un point de jonction. *C.R. Acad. Sci. Paris 234*, 181 (1952).
59. Tan, H. S. *On Mach Reflection and Strength of a Reflected Shock*. Cornell Univ. Grad School of Aero. Engrg., Ithaca. Also *WADC Technical Report 52-163*, Wright Air Dev. Center, Dayton, O., 1952.
60. Miles, J. W. On virtual mass and transient motion in subsonic compressible flow. *Quart. J. Mech. and Appl. Math. 4*, 388 (1951).
61. Longhorn, A. L. The unsteady, subsonic motion of a sphere in a compressible inviscid fluid. *Quart. J. Mech. and Appl. Math. 5*, 64 (1952).
62. Longhorn, A. L. Subsonic compressible flow past bluff bodies. To be published in *Aero. Quart.*
63. Holder, D. W., and Chinneck, A. Observations of the flow past elliptic-nosed two-dimensional cylinders and bodies of revolution in supersonic airstreams. *ARC Report 14216* (unpublished). To be published in *Aero. Quart.*
64. Mair, W. A., and Bardsley, O. Separation of the boundary layer at a slightly blunt edge in supersonic flow. *Phil. Mag. 43*, 344 (1952).
65. Bardsley, O. The conditions at a sharp leading edge in supersonic flow. *Phil. Mag. 42*, 255 (1951).
66. Whitham, G. B. The flow pattern of a supersonic projectile. *Comm. on Pure and Appl. Math. 5*, 301 (1952).
67. Eggers, A. J., and Syvertson, C. A. Inviscid flow about airfoils at high supersonic speeds. *NACA Technical Note 2646*, 1952.

Lightning Source UK Ltd.
Milton Keynes UK
UKHW021529080522
402613UK00005B/279

9 780691 626017